CW01163166

Progress in Inflammation Research

Series Editor

Prof. Dr. Michael J. Parnham
Senior Scientific Advisor
PLIVA dd
Prilaz baruna Filipovica 25
10000 Zagreb
Croatia

Advisory Board

G. Z. Feuerstein (Merck Research Laboratories, West Point, PA, USA
M. Pairet (Boehringer Ingelheim Pharma KG, Biberach a. d. Riss, Germany)
W. van Eden (Universiteit Utrecht, Utrecht, The Netherlands)

Forthcoming titles:
Cancer and Inflammation, D.W. Morgan, M. Nakada, U. Forssmann (Editors), 2004
Anti-Inflammatory or Anti-Rheumatic Drugs,
    R.O. Day, D.E. Furst, P.L. van Riel (Editors), 2004
Cytokines and Joint Injury, P. Miossec, W.B. van den Berg (Editors), 2004
Antibiotics as Antiinflammatory and Immunomodulatory Agents,
    B. Rubin, J. Tamaoki (Editors), 2004

(Already published titles see last page.)

Recent Advances in the Pathophysiology of COPD

T. T. Hansel
P. J. Barnes

Editors

Birkhäuser Verlag
Basel · Boston · Berlin

Editors

Trevor T. Hansel
NHLI Clinical Studies Unit
Royal Brompton Hospital
Fulham Road
London SW3 6HP
United Kingdom

Peter J. Barnes
Department of Thoracic Medicine
National Heart and Lung Institute
Imperial College
Dovehouse St.
London SW3 6LY
United Kingdom

A CIP catalogue record for this book is available from the Library of Congress, Washington D.C., USA

Bibliographic information published by Die Deutsche Bibliothek
Die Deutsche Bibliothek lists this publication in the Deutsche Nationalbibliografie;
detailed bibliographic data is available in the internet at http://dnb.ddb.de

The publisher and editor can give no guarantee for the information on drug dosage and administration contained in this publication. The respective user must check its accuracy by consulting other sources of reference in each individual case.

The use of registered names, trademarks etc. in this publication, even if not identified as such, does not imply that they are exempt from the relevant protective laws and regulations or free for general use.

ISBN 3-7643-6914-0 Birkhäuser Verlag, Basel – Boston – Berlin

This work is subject to copyright. All rights are reserved, whether the whole or part of the material is concerned, specifically the rights of translation, reprinting, re-use of illustrations, recitation, broadcasting, reproduction on microfilms or in other ways, and storage in data banks. For any kind of use, permission of the copyright owner must be obtained.

© 2004 Birkhäuser Verlag, P.O. Box 133, CH-4010 Basel, Switzerland
Part of Springer Science+Business Media
Printed on acid-free paper produced from chlorine-free pulp. TCF ∞
Cover design: Markus Etterich, Basel
Cover illustration: Centrilobular emphysema (see p. 51)
Printed in Germany
ISBN 3-7643-6914-0

9 8 7 6 5 4 3 2 1   www.birkhauser.ch

Contents

List of contributors vii

Jian-Qing He, Scott J. Tebbutt and Peter D. Paré
Genetics of COPD 1

Simonetta Baraldo, Renzo Zuin and Marina Saetta
The pathology of COPD 21

Frances Gilchrist, Onn Min Kon and Michael I. Polkey
Lung function in COPD 31

Rachel C. Tennant, Trevor T. Hansel and David M. Hansell
Computed tomography (CT) scans in COPD 47

Peter J. Barnes
Oxidative stress in COPD 61

Anita L. Sullivan and Robert A. Stockley
Proteinases in COPD 75

Duncan F. Rogers
Mucus hypersecretion in COPD 101

Vera M. Keatings and Clare M. O'Connor
Induced sputum and BAL analysis in COPD 121

Sergei A. Kharitonov and Peter J. Barnes
Exhaled breath markers in COPD 137

Alvar G.N. Agustí
Systemic features of COPD 155

Annemie M.W.J. Schols and Emiel F.M. Wouters
Pulmonary rehabilitation 167

Trevor T. Hansel, Rachel C. Tennant, Edward M. Erin, Andrew J. Tan and Peter J. Barnes
New drugs for COPD based on advances in pathophysiology 189

Index 227

List of contributors

Alvar G.N. Agustí, Servei de Pneumología, Hospital Universitari Son Dureta, IUNICS, Andrea Doria 55, 07014 Palma, Mallorca, Spain; e-mail: aagusti@hsd.es

Simonetta Baraldo, Department of Clinical and Experimental Medicine, Section of Respiratory Diseases, University of Padova, Via Giustiniani 3, 35128 Padova, Italy; e-mail: simonetta.baraldo@unipd.it

Peter J. Barnes, National Heart and Lung Institute (NHLI), Department of Thoracic Medicine, Imperial College School of Medicine, Dovehouse Street, London SW3 6LY, UK; e-mail: p.j.barnes@imperial.ac.uk

Edward M. Erin, National Heart and Lung Institute (NHLI) Clinical Studies Unit, Royal Brompton Hospital, Fulham Road, London SW3 6HP, UK

Frances Gilchrist, Department of Respiratory Medicine, St. Mary's Hospital, Praed Street, London W2 1NY, UK

Trevor T. Hansel, National Heart and Lung Institute (NHLI) Clinical Studies Unit, Royal Brompton Hospital, Fulham Road, London SW3 6HP, UK; e-mail: t.hansel@imperial.ac.uk

David M. Hansell, Department of Radiology, Royal Brompton Hospital, Fulham Road, London SW3 6HP, UK; e-mail: d.hansell@rbh.nthames.nhs.uk

Jian-Qing He, University of British Columbia, James Hogg iCAPTURE Center for Cardiopulmonary Research, St. Paul's Hospital, Burrard Building, 1081 Burrard Street, Vancouver, B.C. V6Z 1Y6, Canada

Sergei A. Kharitonov, Department of Thoracic Medicine, National Heart and Lung Institute, Imperial College London, Royal Brompton Hospital, Dovehouse Street, London SW3 6LY, UK; e-mail: s.kharitonov@imperial.ac.uk

Vera M. Keatings, Department of Respiratory Medicine, Letterkenny General Hospital, Letterkenny, Co. Donegal, Ireland; e-mail: vera.keatings@nwhb.ie

List of contributors

Onn Min Kon, Department of Respiratory Medicine, St. Mary's Hospital, Praed Street, London W2 1NY, UK

Clare M. O'Connor, The Conway Institute for Biomolecular and Biomedical Research and the Dublin Molecular Medical Centre, University College Dublin, Belfield, Dublin 4, Ireland; e-mail: clare.oconnor@ucd.ie

Peter D. Paré, University of British Columbia, James Hogg iCAPTURE Center for Cardiopulmonary Research, St. Paul's Hospital, Burrard Building, 1081 Burrard Street, Vancouver, B.C. V6Z 1Y6, Canada; e-mail: ppare@mrl.ubc.ca

Michael I. Polkey, Royal Brompton Hospital & National Heart and Lung Institute, Fulham Road, London SW3 6NP, UK; e-mail: m.polkey@rbh.nthames.nhs.uk

Duncan F. Rogers, Thoracic Medicine, National Heart & Lung Institute, Imperial College London, Dovehouse Street, London SW3 6LY, UK; e-mail: duncan.rogers@imperial.ac.uk

Marina Saetta, Department of Clinical and Experimental Medicine, Section of Respiratory Diseases, University of Padova, Via Giustiniani 3, 35128 Padova, Italy; e-mail: marina.saetta@unipd.it

Annemie M.W.J. Schols, Department of Respiratory Medicine, University Hospital Maastricht, P.O. Box 5800, 6202 AZ Maastricht, The Netherlands; e-mail: a.schols@pul.unimaas.nl

Robert A. Stockley, Department of Respiratory Medicine, 1st Floor Nuffield House, Queen Elizabeth Hospital, Edgbaston, Birmingham B15 2TH, UK; e-mail: r.a.stockley@bham.ac.uk

Anita L. Sullivan, Department of Respiratory Medicine, 1st Floor Nuffield House, Queen Elizabeth Hospital, Edgbaston, Birmingham B15 2TH, UK; e-mail: a.l.sullivan@bham.ac.uk

Andrew J. Tan, National Heart and Lung Institute (NHLI) Clinical Studies Unit, Royal Brompton Hospital, Fulham Road, London SW3 6HP, UK; e-mail: a.tan@imperial.ac.uk

Scott J. Tebbutt, University of British Columbia, James Hogg iCAPTURE Center for Cardiopulmonary Research, St. Paul's Hospital, Burrard Building, 1081 Burrard Street, Vancouver, B.C. V6Z 1Y6, Canada

Rachel C. Tennant, National Heart and Lung Institute (NHLI) Clinical Studies Unit, Royal Brompton Hospital, Fulham Road, London SW3 6HP, UK;
e-mail: r.tennant@imperial.ac.uk

Emiel F.M. Wouters, Department of Respiratory Medicine, University Hospital Maastricht, P.O. Box 5800, 6202 AZ Maastricht, The Netherlands;
e-mail: e.wouters@lung.azm.nl

Renzo Zuin, Department of Clinical and Experimental Medicine, Section of Respiratory Diseases, University of Padova, Via Giustiniani 3, 35128 Padova, Italy;
e-mail: renzo.zuin@unipd.it

Genetics of COPD

Jian-Qing He, Scott J. Tebbutt, Peter D. Paré

University of British Columbia, James Hogg iCAPTURE Center for Cardiopulmonary Research, St. Paul's Hospital, Burrard Building, 1081 Burrard St, Vancouver, B.C., V6Z 1Y6 Canada

Introduction

The prevalence of COPD is increasing worldwide and has become the fourth leading cause of death in the United States. Although cigarette smoking is by far the most important risk factor, only 10–20% of smokers develop symptomatic COPD, and less than 15% of the variation in lung function among smokers can be explained by the extent and duration of cigarette smoking. These data indicate that host and/or environmental factors other than simple exposure must contribute to disease pathogenesis. Family and twin studies suggest that at least some of this variance is genetic. In this chapter we discuss approaches to identify disease-causing genes and provide a current summary of the results of linkage and association studies.

COPD phenotypes

A critical step in identifying disease-causing genes is the definition of phenotype. For years, clinicians, physiologists, pathologists and epidemiologists have debated the definition of COPD. In Table 1 we briefly summarize the phenotypes that have been, and/or could be used in the search for genes. Among them, the most common phenotypes for COPD genetic studies are the presence and degree of airflow obstruction and its rate of change over time. Airflow obstruction can occur on the basis of either of two very different pathophysiological processes in the lung: 1) inflammation of the parenchyma resulting in proteolysis of the lung parenchyma and loss of lung elasticity (emphysema); and 2) inflammation, scarring and narrowing of the small airways ("small airway disease"). In an individual patient one of these

Table 1 - COPD phenotypes in genetic studies

Cross-sectional categorical and quantitative phenotypes	
Chronic bronchitis	Defined by a productive cough of more than three months duration for more than two successive years; this reflects mucous hypersecretion and is only weakly related to airflow obstruction
Emphysema	Defined as permanent enlargement of air spaces accompanied by destruction of the lung parenchyma. This lesion can be measured pathologically or by CT scanning and causes loss of lung recoil, hyperinflation and gas trapping.
COPD	Defined functionally, usually on the basis of reduced FEV_1 % predicted and FEV_1/FVC ratio. No universally accepted cut off values are available but the recently proposed GOLD categories have offered a new approach to standardization.
Lung function	Various measures of lung function can be used as quantitative traits in genetic studies

Longitudinal quantitative phenotypes	
Rate of decline in lung function	Since accelerated decline in lung function in response to cigarette smoking is believed to be the most important pathophysiological event leading to COPD, this is a potentially very powerful phenotype
Rate of lung growth	Since maximally achieved lung function is thought to be one potential source of variable response to cigarette smoke, genetic and environmental factors which influence lung and airway development are potentially important factors
Age at onset of decline of lung function	Since one loses lung function as one age, the earlier this process starts, the more likely lung function will reach a threshold that will lead to symptoms

processes, which may be controlled by different genetic factors, may predominate although both usually co-exist. Both of these processes ultimately produce similar patterns of functional impairment: decreased expiratory flow, hyperinflation and abnormalities of gas exchange. Until recently it has been impossible to separate these two processes. However, the recent demonstration by Nakano et al. [1] that the relative contribution of airway and parenchymal disease can be separated in smokers who have COPD by using high resolution computed tomography (HRCT), suggests that they should be measured and analyzed separately in future genetic studies.

Approaches to identify susceptibility genes

The two main strategies used to identify susceptibility genes are genomic scans and association studies.

Genomic scans

This approach involves searching the entire genome for regions that harbor disease-causing genes by linkage analysis. It usually requires affected families of at least two generations. Each family member is typed for DNA markers that are scattered throughout the genome. Linkage analysis determines whether any of the markers are inherited with the disease more than predicted by chance. If so, the next step is to perform fine mapping of the region with the aim of identifying novel or candidate genes near the linked marker(s). The advantages of a genomic scan are that novel genes can be identified for the pathogenesis of a disease and that it is immune to confounding due to population admixture (see below). A good example of this approach was shown in the finding of the *ADAM33* gene for asthma [2]. However, the disadvantage is the requirement for families with several exposed and affected members in whom accurate phenotypic data are available. This requirement makes genetic dissection of diseases with a late age of onset, such as COPD, difficult.

Only two groups have reported linkage analysis in COPD [3, 4]. Silverman et al. enrolled 72 individuals who had severe, early-onset COPD and 585 of their relatives. Using qualitative phenotypes of airway obstruction (mild and moderate COPD) and chronic bronchitis they found suggestive evidence for linkage (LOD score > 1.21) on chromosomes 8, 12, 19, and 22. The highest two-point LOD score (3.14) was seen when the analysis was restricted to smokers and when mild obstruction was the phenotype. For the quantitative phenotypes, FEV_1, FVC and FEV_1/FVC, multipoint variance-component linkage analysis was performed and the highest LOD score was 4.12 for FEV_1/FVC on Chromosome 2q. The highest LOD score for FEV_1 was 2.43 on chromosome 12 in the same region that was linked in the qualitative study.

Joost et al. [5] used markers spaced at ~10 cM in a genome-wide scan of 1,578 members of 330 families participating in the Framingham Study to test for linkage of genetic markers to level of lung function, as determined by spirometry during middle age (48–55 years). Lung function at this age reflects lung growth, the maximal lung function achieved and the rate of lung function decline. After correction for age and smoking status they found evidence of linkage for FEV_1 and FVC on chromosomes 4, 6 and 21 (maximum log score for FEV_1 was 2.4 on chromosome 6 and for FVC was 2.6 on chromosome 21).

Interestingly, there is no co-localization of the loci identified within these two studies by Silverman et al. and Joost et al. Whether this can be explained by the differences in phenotypes that were studied or by the vagaries of chance remains to be determined. However, because both studies use DNA markers with an average spacing of 9–10 cM, there is a risk of missing important sites of linkage, since linkage disequilibrium is unlikely to extend over more than 1–2 cM [6]. A set of 2,000 or more DNA markers would increase the resolution of genome scans. Another potential limitation of these studies is that genome scans in larger extended families may give useful information about genes important in those families but may not be relevant to the population at large.

The affected sib-pair method is the other way to test, genome-wide, for linkage and it is more suitable for late onset complex diseases such as COPD. Using this method, greater than expected sharing of alleles at a locus by sibs who smoke and are affected (i.e., develop airflow obstruction) suggests that the locus contains, or is near, a disease-causing gene. To date there have been no reports of such studies in COPD. Large sample size will be required to find genes of moderate effect in a disease with relatively low heritability such as COPD. A multi-center consortium is in the process of identifying ~1,000 affected sib pairs with COPD [7]. It is anticipated that more information will be available from similar studies within the next five years.

Association studies

This approach involves choosing candidate genes that are implicated in the pathogenesis of the disease. A disadvantage of association studies is that only known genes can be examined. There are several criteria to be considered in choosing the most likely candidate genes [8]. Firstly, the gene product must be a protein likely to be relevant to disease pathophysiology. Secondly, genes which lie within, or close to, linked loci should be given priority. Finally, the gene must contain single nucleotide polymorphisms (SNPs) or other types of genetic variation capable of producing functional differences in either the level of expression or the function of the gene product (most probably within coding or regulatory regions). Although the design and implementation of association studies are relatively straightforward there are a number of potential sources of error, and large sample size and relatively high allele frequency are necessary to power studies sufficiently, since each gene polymorphism is likely to have a small effect.

The two basic designs of association studies are cohort studies and case-control studies. To date, most association studies in COPD have been case-control studies, which involve unrelated individuals. Although this approach circumvents many of the problems associated with family studies there is a risk of false positive results due to population admixture.

Population admixture can result in false positive and false negative association. For example, a false positive result could come about if one ethnic group had a higher incidence of COPD (based on unmeasured environmental factors) and a different frequency distribution of candidate SNPs was admixed within the case group. Any SNP more common in this ethnic group would be spuriously "associated" with the phenotype. This pitfall emphasizes the importance of careful selection of cases and controls. Recently, several methods have been advocated to attempt to correct for population admixture. One of these approaches is the use of genomic controls [9, 10]. This strategy measures the degree of statistically identified genotype-phenotype association for a large number (60 or more) of irrelevant SNPs. The association of a disease with irrelevant SNP suggests that there is a population subset with different genetic backgrounds plus different disease susceptibility. Such a spurious association can then be corrected in order to derive true gene-disease associations [11, 12].

Another potential hazard of association studies is the tendency to test multiple polymorphisms in many genes using a number of phenotypes. Multiple comparisons inevitably result in false positive associations. This tendency, coupled with the bias resulting from failure to publish negative studies, is likely to be the basis of the frequent observation that association studies cannot be replicated. The expected frequency of false positives is given by $1-(1-k)^m$ (where m is the number of independent markers and k is usually 0.05, the significance level set for a single marker) [13]. An additional source of uncertainty in some of these studies is related to apparent association between a genetic variant and a disease which may be explained by other confounding factors such as environmental exposure [14]. Finally, linkage disequilibrium within the genome makes it difficult to implicate specific SNPs in disease pathogenesis. Linkage disequilibrium is the tendency of SNPs to be grouped together as haplotypes on chromosomes and is related to the low probability that chromosomal recombination will separate two SNPs that are in close proximity on a chromosome over the relatively brief period of human evolution. A polymorphic marker that is associated with a disease phenotype may be in linkage disequilibrium with another allele at a nearby locus that is the true susceptibility allele.

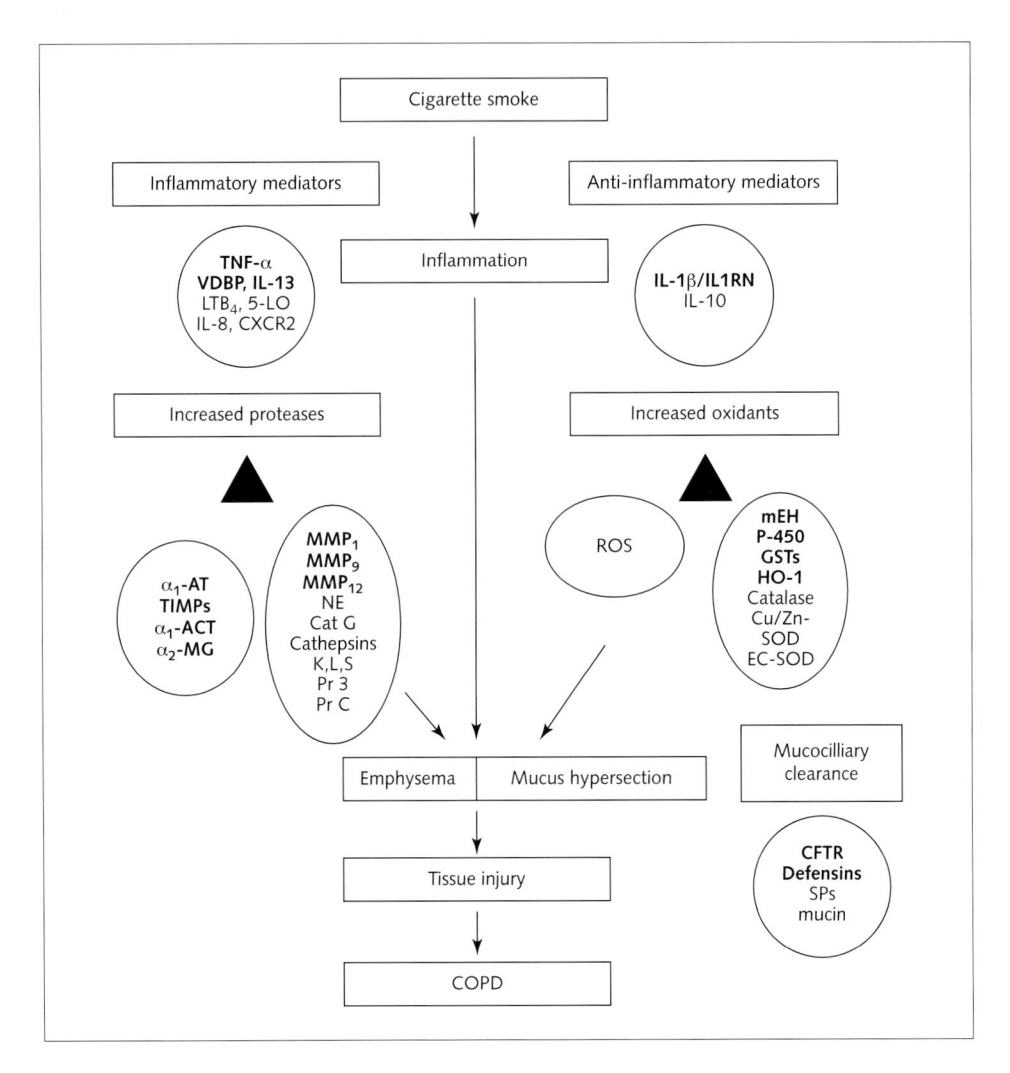

Figure 1
Summary of pathways and possible candidate genes involved in the pathogenesis of COPD. Genes that have been associated with COPD are in bold font. (ROS, reactive oxygen species; TNF-α, tumor necrosis factor-α; VDBP, vitamin D-binding protein; IL-1β/IL1RN, interleukin-1β/interleukin-1β receptor-antagonist; LTB$_4$, leukotriene B$_4$; 5-LO, 5-lipoxygenase; α_1-AT, α_1-antitrypsin; TIMP, tissue inhibitors of metalloproteinases; α_1-ACT, α_1-antichymotrypsin; α_2-MG, α_2-macroglobulin; MMPs, matrix metalloproteinases; NE, neutrophil elastase; CatG, cathepsin G; Pr3, proteinase 3; mEH, microsomal epoxide hydrolase; P450, cytochrome P-450; GST, glutathione S-transferase; HO-1, heme oxygenase-1; Cu/Zn-SOD, copper-zinc superoxide dismutase; EC-SOD, extracellular SOD; CFTR, cystic fibrosis transmembrane regulator; SPs, surfactant proteins).

Candidate genes studied in COPD

The protease-antiprotease and oxidant-antioxidant hypotheses have dominated our thinking in the pathogenesis of smoke-related COPD. Both hypotheses are closely related to modulation of the consequences of the inflammatory process. The genes involved, or potentially involved, in the pathogenesis of COPD are summarized in Figure 1.

Protease and antiprotease genes

α_1-antitrypsin (α_1-AT)

α_1-AT deficiency is the most convincing and well characterized genetic risk factor for COPD [15]. The α_1-AT is the major plasma protease inhibitor of neutrophil elastase. The PI locus is polymorphic; in the Caucasian population, the frequencies of M, S and Z alleles are > 95%, 2–3% and 1% respectively. These alleles are associated with normal, mildly-reduced and severely-reduced α_1-AT levels. A small percentage of people inherit a null allele, which leads to complete absence of α_1-AT production. Individuals with two Z alleles or one Z and one null allele are referred as PI Z. PI Z individuals have approximately 15% of normal plasma α_1-AT levels. The levels are low because 85% of the synthesized mutant Z α_1-AT is retained as polymers within hepatocytes. Individuals with MS and MZ genotypes have ~80% and 60% of normal α_1-AT levels, respectively. Heterozygous PI SZ is rare and individuals with this genotype have α_1-AT levels ~40% of normal. Some of the association studies of α_1-AT, other antiprotease and protease genes and COPD are shown in Table 2.

Other antiprotease genes

Tissue inhibitors of metalloproteinases (TIMPs)

TIMPs are inhibitors of the matrix metalloproteinases (MMPs). Four members of the TIMP family (TIMP1–4) interact with the active form of MMPs to inhibit their activities. TIMP1 and TIMP2 have the potential to participate in pulmonary diseases such as emphysema. TIMP2 is a more effective inhibitor of MMP2 and MMP9 than MMP1 [31]. There is some evidence to suggest that the membrane-type MMP1 (MT1MMP)/MMP2/TIMP2 system plays a role in the pathogenesis of pulmonary emphysema [32].

α_1-antichymotrypsin (α_1-ACT)

α_1-ACT belongs to the serine protease inhibitor family and is expressed in alveolar macrophages and airway epithelia.

Table 2 - Associations of protease-antiprotease gene polymorphisms and COPD

Antiprotease genes

1. α_1-antitypsin (α_1-AT)
- ZZ genotype: associated with emphysema [16]; accelerated rate of decline of lung function [17]; early-onset of COPD [16, 18]
- MS and MZ genotypes: MZ genotype frequency increased in COPD, MS genotype not consistent [19, 20]
- SZ genotype: increase prevalence of COPD [21]
- 3' Taq-1 polymorphism: increased frequency in emphysema [22] and COPD [23]; not associated with COPD [24]
- 3' Hind-III polymorphism: Increased frequency in COPD [25]
2. TIMP-2: associated with COPD [26]
3. α_1-ACT: associated with COPD [27]; not associated with COPD [24]
4. α_2-MG: associated with COPD [28]

Protease genes

1. MMP1: G-1607GG: associated with rapid decline of lung function [29]
2. MMP9:
- –1562C/T: associated with emphysema [30]; not associated with decline of lung function [29]
- CA repeat: not associated with decline of lung function [29]
3. MMP12: Asn357Ser: not associated with decline of lung function [29]

Proteinase genes

Matrix metalloproteinases (MMPs)

MMPs comprise a structurally- and functionally-related family of at least 20 proteolytic enzymes that play essential roles in tissue remodeling. Over-expression of human *MMP1* results in emphysema in transgenic mice, and deletion of *MMP12* in mouse models prevents smoking-related emphysema. MMPs have been shown to play a role in the pathogenesis of pulmonary emphysema in humans. The insertion of a G at position –1607 (–1607GG) in the promoter of *MMP1* gene was found to be associated with a higher levels of gene expression [33]. Two polymorphisms in the promoter region of *MMP9* have been associated with promoter activity, with higher level of expression for the T allele in the –1562C/T polymorphism or smaller number of CA repeats [34, 35].

Antioxidant and xenobiotic metabolizing enzyme genes

Heme oxygenase-1 (HMOX1)
HMOX1 is a key enzyme in heme catabolism, functioning as an antioxidant enzyme. The *HMOX1* gene (GT)n dinucleotide repeat in the 5'-flanking region shows sequence length polymorphism and specific alleles are associated with different levels of gene transcription during thermal stress [36]. There is also *in vitro* evidence that a high number of (GT)n repeats may reduce HMOX1 inducibility by reactive oxygen species (ROS) in cigarette smoke [37].

The results of association studies of *HMOX1* and xenobiotic metabolizing enzyme genes are presented in Table 3.

Xenobiotic metabolizing enzyme genes

Microsomal epoxide hydrolase (mEH)
mEH metabolizes polycyclic aromatic hydrocarbons, which are carcinogens found in cigarette smoke. Two well-characterized polymorphisms have been described: the first, within exon 3, results in the substitution of Tyr with His at amino acid position 113; another, within exon 4, codes for the substitution of a His with Arg at amino acid position 139. The change from 113Tyr to 113His causes reduced enzyme activity to at least 50% (slow allele), and 139His to 139Arg causes increased enzyme activity by at least 25% (fast allele).

Glutathione-S-transferases (GST)
GSTs are a family of enzymes that appear to be critical in protection against oxidative stress by detoxifying various toxic substrates in tobacco smoke. Several GST polymorphisms have been identified. *GSTM1* has three alleles: *GSTM1*0* is a null allele in which the entire gene is absent while *GSTM1*A* and *GSTM1*B* encode monomers that form active enzymes. Obviously, homozygosity for the *GSTM1* null allele results in a complete lack of GSTM1 activity. *GSTT1* has a similar null genotype. GSTP1 contains two polymorphisms: an A→G transition at nucleotide +313 that leads to the 105Ile/Val substitution and a C→T transition at nucleotide +341 that leads to the 114Ala/Val substitution. The 105 substitution has been shown to result in altered catalytic activity [59], though there is currently no evidence of a functional effect of the 114 substitution.

Cytochrome P4501A1
The polycyclic aromatic hydrocarbon-metabolizing cytochrome P450 isoform, P4501A1 (CYP1A1), is an enzyme that metabolizes exogenous compounds to

Table 3 - Association studies of antioxidant and inflammatory mediator gene polymorphisms and COPD

Gene/genotype	Phenotype of COPD	Association	Population	Refs
Heme oxygenase-1 (HMOX1): GT repeat	Emphysema	Yes	Japanese	[37]
	Rate of lung function ↓	No	Whites	[38]
Microsomal epoxide hydrolase (mEPHX): 113His/His139	Emphysema	Yes	Caucasians	[39]
	COPD	Yes	Caucasians	[39]
	COPD	No	Koreans	[40]
	Rate of lung function ↓	Yes	Whites	[19]
GSTM1 null	Chronic bronchitis	Yes	Caucasians	[41]
	Emphysema with cancer	Yes	Caucasians	[42]
	COPD	No	Koreans	[40]
	Lung function growth	Yes	Whites	[43]
	Rate of lung function ↓	No	Whites	[38]
GSTT1 null	COPD	No	Koreans	[40]
	Lung function growth	No	Whites	[43]
	Rate of lung function ↓	No	Whites	[38]
GSTP1 105Ile	COPD	Yes	Japanese	[44]
	COPD	No	Koreans	[45]
	Rate of lung function ↓	No	Whites	[38]
105Val	Lung function growth	Yes	Whites	[43]
Cytochrome P4501A1: 462Val	Emphysema with cancer	Yes	Caucasians	[46]
TNF-α: −308A	Chronic bronchitis	Yes	Chinese	[47]
	Emphysematous change (HRCT)	Yes	Japanese	[48]
	COPD	Yes		[49]
	COPD/Airflow obstruction	No/Yes	Whites	[50]
	COPD	No	Japanese	[51]
	COPD/Airflow obstruction	No/No	Caucasians	[52]
	Rate of lung function ↓	No	Whites	[19]
−308A/G or −376G/A or −238G/A	COPD/emphysema	No/No	Dutch	[53]
+489G/A	COPD/emphysema	Yes/Yes	Dutch	[53]

Table 3 (continued)

Gene/genotype	Phenotype of COPD	Association	Population	Refs
Vitamin D binding protein	COPD	Yes	Caucasians	[54]
	COPD	Yes	Japanese	[55]
	Rate of lung function ↓	No	Whites	[19]
IL-1β: –511C/T	COPD	No	Japanese	[51]
	Rate of lung function ↓	No	Whites	[56]
IL1RN: 86bp Tandem repeat	COPD	No	Japanese	[51]
	Rate of lung function ↓	No	Whites	[56]
IL-1β/IL1RN haplotypes	Rate of lung function ↓	Yes	Whites	[56]
IL-13: –1055C/T	COPD	Yes	Whites	[57]
	Rate of lung function ↓	No	Whites	[58]
Arg130Gln	COPD	No	Whites	[57]
	Rate of lung function ↓	No	Whites	[58]

enable them to be excreted in the urine or bile. The most common allelic variants of *CYP1A1* are the *Msp*I restriction site located in the 3'-flanking region, the Ile462Val polymorphism located in exon 7, and the C-459T polymorphism located in the promoter. The 462Val isoform results in increased CYP1A1 activity *in vivo* [60].

Inflammatory mediator genes

Tumor necrosis factor-alpha (TNF-α)

TNF-α is a pro-inflammatory cytokine, whose production is elevated in the airways of COPD patients, especially during acute exacerbations [61]. The G-308A polymorphism in the promoter of *TNF-α* has been associated with higher expression of TNF-α [62].

The associations of SNPs in *TNF-α* and other inflammatory mediator genes with COPD are listed in Table 3.

Vitamin D binding protein (VDBP)

VDBP has effects on neutrophil chemotaxis and macrophage activation. Two common SNPs that have been extensively studied cause amino acid variants: a GAT to

GAG substitution at codon 416 causes replacement of aspartic acid by glutamic acid and the ACG to AAG substitution at codon 420 changes threonine to lysine. Only three of the possible four isoforms formed by combinations of these amino acid substitutions exist due to linkage disequilibrium and these isoforms are named 1F, 1S, and 2 [63].

Interleukin-1 (IL-I) complex
The IL-1 family consists of pro-inflammatory cytokines (IL-1α, IL-1β) and the anti-inflammatory agent, IL-1 receptor antagonist (IL1RN). IL1RN competitively inhibits the binding of IL-1α and IL-1β to its receptor but does not induce any intracellular response. A base pair substitution at position +4845 from the transcription start site in exon 5 of the IL-1α gene results in an 114Ala to 114Ser substitution. There are two SNPs in the IL-1β gene: C-551T in the promoter region and C3953T in exon 5. The gene encoding IL1RN has a six-allelic 86-bp tandem repeat in intron 2. Evidence of gene-gene interaction at these loci includes the enhancing effect of the IL-1RN allele 2 (containing two repeats) on IL-1RA plasma levels requiring the presence of the IL-1β −511T or the absence of the IL-1β +3953T [64].

Interleukin-13 (IL-13)
Targeted expression of IL-13 in the adult murine lung has been shown to cause emphysema. Two SNPs of Arg130Gln and C-1112T have been associated with differences in the plasma level of IL-13 and asthma phenotype.

Genes involved in airway defense

Cystic fibrosis transmembrane conductance regulator (CFTR)
The CFTR protein forms a chloride channel at the apical surface of airway epithelial cells and is involved in the control of airway secretions. Over 1,000 mutations have been identified in the gene so far and they are classified into five classes based on the molecular alteration at the protein level [65]. Class 1 and class 2 mutations lead to complete absence, or an altered localization, respectively, of the CFTR protein, thereby having the most dramatic effect on function: ΔF508, the most frequent of these disease-causing mutations, is an example of a class 2 mutation. Class 3 mutations affect the regulatory domains of the CFTR protein or lead to production of CFTR with reduced chloride transport. Class 4 mutations reduce the ionic conductance of the chloride channel, and class 5 mutations lead to a reduced level of expression of a normal CFTR protein. IVS8, which is an example of a class 5 mutation, is a variable length thymine repeat in intron 8 of the *CFTR* gene.

Heterozygosity of ΔF508 has been found to be increased in patients who have disseminated bronchiectasis [66], and in patients who have "bronchial hypersecretion" [67]. The prevalence of ΔF508 was not increased in patients who had chronic bronchitis [66]. Studies of IVS8-5T as a risk factor for COPD have yielded conflicting results [68, 69]. In one study, the IVS8-5T allele was found in one out of 12 COPD patients compared with one out of 52 controls, which the authors report as a significant difference. The frequency of the Met allele of the Met470Val polymorphism was increased in patients (71%) compared with controls (36%) [70]. However, the Met470 variant is associated with increased CFTR chloride channel activity compared with the Val variant [71] and therefore the reason for the association with COPD remains unclear.

Defensins

Defensins are small cationic peptides produced by inflammatory and structural tissue cells. They have broad spectrum antimicrobial activity and act as key effector molecules in innate immunity. SNPs in *β-defensin-1* (*hBD-1*) were investigated in 60 COPD patients and 213 controls; a genotype containing the 38Ile allele was observed in 15% of patients but only 2.8% of controls (OR = 6.1, 96% CI = 2.0–18.3, p = 0.0012) [72].

Surfactant proteins (SPs)

In addition to their role as modulators of lung surface tension, SPs play an important role in innate host defense. In one study the investigators analyzed the association of SP-A, SP-B, and SP-D SNPs as well as a SP-B-linked microsatellite marker with COPD. Several COPD susceptibility alleles (SP-A codon 62 A allele, SP-B nucleotide 1508C and SP-B-linked microsatellite D2S388_5) were identified as possible susceptibility loci [73]. However, these observations need to be confirmed in larger studies.

Mucin

Mucus, derived from goblet cells and airway submucosal glands, covers the epithelial surface of the respiratory tract, preventing desiccation and providing lubrication and defense from bacteria and harmful enzymes. The MUC genes are highly variable due to different lengths of repeat sequences which code for the peptide anchor points for highly branched hydroscopic sugar molecules. The longer the repeat sequences the longer the MUC proteins and the larger the eventual mucous glycoprotein. Of the 12 mucin genes that have been characterized, longer MUC2 and MUC7*5 repeat alleles have been found to confer protection against the development of asthma [74, 75]. However, no study has been done in COPD.

The future of genetics of COPD

In this chapter, we have reviewed methods of identifying susceptibility genes in complex diseases such as COPD, and briefly summarized the data implicating known or putative susceptibility genes for this disorder. The bulk of the studies have been case-control association studies in which only known genes can be tested. The results of these association studies have identified heterozygosity for α_1-AT, slow variants of microsomal epoxide hydrolase and deletion variants of glutathione transferases as likely susceptibility genes. However, many more candidate genes have been inconsistently associated with COPD and their role in pathogenesis awaits larger and more powerful study designs. In the majority of investigations, only individual SNPs have been studied in and around a single candidate gene. Negative results do not rule out an association involving other nearby SNPs. Positive results do not mean the discovery of the causal SNP since the result may reflect linkage disequilibrium (LD) with a true causal SNP located some distance away [76]. Future genetic studies of COPD should focus on large, well characterized cohort and case control designs, and especially on the performance of family or affected sib-pair analyses. Although it is difficult to perform family or affected sib-pair analysis in COPD, these studies do have the potential to identify novel disease-causing genes and to overcome the problems associated with population admixture. When the possibility of population admixture exists, genomic controls or similar methods should be used in cohort and case-control studies to avoid false results. Much more precision of gene identification will be achieved by using more robust COPD phenotypes, including the separation of patients who have airway-predominant and parenchymal-predominant lesions. The use of large haplotype blocks of SNPs that are in linkage disequilibrium will allow more effective genome-wide association studies [77]. Finally, sophisticated study designs will allow the consideration of gene–gene and gene–environment interactions that are absolutely necessary to explore the complex pathophysiology of chronic obstructive pulmonary disease.

References

1 Nakano Y, Muro S, Sakai H, Hirai T, Chin K, Tsukino M, Nishimura K, Itoh H, Pare PD, Hogg JC et al (2000) Computed tomographic measurements of airway dimensions and emphysema in smokers. Correlation with lung function. *Am J Respir Crit Care Med* 162: 1102–1108
2 Van Eerdewegh P, Little RD, Dupuis J, Del Mastro RG, Falls K, Simon J, Torrey D, Pandit S, McKenny J, Braunschweiger K et al (2002) Association of the ADAM33 gene with asthma and bronchial hyperresponsiveness. *Nature* 418: 426–430
3 Silverman EK, Mosley JD, Palmer LJ, Barth M, Senter JM, Brown A, Drazen JM, Kwiatkowski DJ, Chapman HA, Campbell EJ et al (2002) Genome-wide linkage analy-

sis of severe, early-onset chronic obstructive pulmonary disease: airflow obstruction and chronic bronchitis phenotypes. *Hum Mol Genet* 11: 623–632

4   Silverman EK, Palmer LJ, Mosley JD, Barth M, Senter JM, Brown A, Drazen JM, Kwiatkowski DJ, Chapman HA, Campbell EJ et al (2002) Genome-wide linkage analysis of quantitative spirometric phenotypes in severe early-onset chronic obstructive pulmonary disease. *Am J Hum Genet* 70: 1229–1239

5   Joost O, Wilk JB, Cupples LA, Harmon M, Shearman AM, Baldwin CT, O'Connor GT, Myers RH, Gottlieb DJ (2002) Genetic loci influencing lung function: a genome-wide scan in the Framingham Study. *Am J Respir Crit Care Med* 165: 795–799

6   Ober C, Cox NJ, Abney M, Di Rienzo A, Lander ES, Changyaleket B, Gidley H, Kurtz B, Lee J, Nance M et al (1998) Genome-wide search for asthma susceptibility loci in a founder population. The Collaborative Study on the Genetics of Asthma. *Hum Mol Genet* 7: 1393–1398

7   Lomas DA, Silverman EK (2001) The genetics of chronic obstructive pulmonary disease. *Respir Res* 2: 20–26

8   Hall IP (1998) Genetic factors in asthma severity. *Clin Exp Allergy* 28 (Suppl 5): 16–20; discussion 6–8

9   Devlin B, Roeder K (1999) Genomic control for association studies. *Biometrics* 55: 997–1004

10 Devlin B, Roeder K, Bacanu SA (2001) Unbiased methods for population-based association studies. *Genet Epidemiol* 21: 273–284

11 Pritchard JK, Rosenberg NA (1999) Use of unlinked genetic markers to detect population stratification in association studies. *Am J Hum Genet* 65: 220–228

12 Bacanu SA, Devlin B, Roeder K (2000) The power of genomic control. *Am J Hum Genet* 66: 1933–1944

13 Cardon LR, Bell JI (2001) Association study designs for complex diseases. *Nat Rev Genet* 2: 91–99

14 Vineis P, McMichael AJ (1998) Bias and confounding in molecular epidemiological studies: special considerations. *Carcinogenesis* 19: 2063–2067

15 Silverman EK (2001) Genetics of chronic obstructive pulmonary disease. *Novartis Found Symp* 234: 45–58; discussion 64

16 Larsson C (1978) Natural history and life expectancy in severe alpha1-antitrypsin deficiency, Pi Z. *Acta Medica Scandinavica* 204: 345–351

17 Piitulainen E, Eriksson S (1999) Decline in FEV_1 related to smoking status in individuals with severe alpha1-antitrypsin deficiency (PiZZ). *Eur Respir J* 13: 247–251

18 Seersholm N, Dirksen A, Kok-Jensen A (1994) Airways obstruction and two year survival in patients with severe alpha 1-antitrypsin deficiency. *Eur Respir J* 7: 1985–1987

19 Sandford AJ, Chagani T, Weir TD, Connett JE, Anthonisen NR, Paré PD (2001) Susceptibility genes for rapid decline of lung function in the Lung Health Study. *Am J Respir Crit Care Med* 163: 469–473

20 Morse JO, Lebowitz MD, Knudson RJ, Burrows B (1977) Relation of protease inhibitor phenotypes to obstructive lung diseases in a community. *N Engl J Med* 296: 1190–1194

21 Turino GM, Barker AF, Brantly ML, Cohen AB, Connelly RP, Crystal RG, Eden E, Schluchter MD, Stoller JK (1996) Clinical features of individuals with PI*SZ phenotype of α_1-antitrypsin deficiency. α_1-Antitrypsin Deficiency Registry Study Group. *Am J Respir Crit Care Med* 154: 1718–1725

22 Kalsheker NA, Hodgson IJ, Watkins GL, White JP, Morrison HM, Stockley RA (1987) Deoxyribonucleic acid (DNA) polymorphism of the α_1-antitrypsin gene in chronic lung disease. *Br Med J* 294: 1511–1514

23 Poller W, Meisen C, Olek K (1990) DNA polymorphisms of the α_1-antitrypsin gene region in patients with chronic obstructive pulmonary disease. *Eur J Clin Invest* 20: 1–7

24 Bentazzo MG, Gile LS, Bombieri C, Malerba G, Massobrio M, Pignatti PF, Luisetti M (1999) α_1-antitrypsin Taq I polymorphism and α_1-antichymotrypsin mutations in patients with obstructive pulmonary disease. *Respir Med* 93: 648–654

25 Kalsheker NA, Watkins GL, Hill S, Morgan K, Stockley RA, Fick RB (1990) Independent mutations in the flanking sequence of the α_1-antitrypsin gene are associated with chronic obstructive airways disease. *Dis Markers* 8: 151–157

26 Hirano K, Sakamoto T, Uchida Y, Morishima Y, Masuyama K, Ishii Y, Nomura A, Ohtsuka M, Sekizawa K (2001) Tissue inhibitor of metalloproteinases-2 gene polymorphisms in chronic obstructive pulmonary disease. *Eur Respir J* 18: 748–752

27 Ishii T, Matsuse T, Teramoto S, Matsui H, Hosoi T, Fukuchi Y, Ouchi Y (2000) Association between alpha-1-antichymotrypsin polymorphism and susceptibility to chronic obstructive pulmonary disease. *Eur J Clin Invest* 30: 543–548

28 Poller W, Barth J, Voss B (1989) Detection of an alteration of the alpha 2-macroglobulin gene in a patient with chronic lung disease and serum alpha 2-macroglobulin deficiency. *Hum Genet* 83: 93–96

29 Joos L, He JQ, Shepherdson MB, Connett JE, Anthonisen NR, Paré PD, Sandford AJ (2002) The role of matrix metalloproteinase polymorphisms in the rate of decline in lung function. *Hum Mol Genet* 11: 569–576

30 Minematsu N, Nakamura H, Tateno H, Nakajima T, Yamaguchi K (2001) Genetic polymorphism in matrix metalloproteinase-9 and pulmonary emphysema. *Biochem Biophys Res Commun* 289: 116–119

31 Howard EW, Bullen EC, Banda MJ (1991) Preferential inhibition of 72- and 92-kDa gelatinases by tissue inhibitor of metalloproteinases-2. *J Biol Chem* 266: 13070–13075

32 Ohnishi K, Takagi M, Kurokawa Y, Satomi S, Konttinen YT (1998) Matrix metalloproteinase-mediated extracellular matrix protein degradation in human pulmonary emphysema. *Lab Invest* 78: 1077–1087

33 Kanamori Y, Matsushima M, Minaguchi T, Kobayashi K, Sagae S, Kudo R, Terakawa N, Nakamura Y (1999) Correlation between expression of the matrix metalloproteinase-1 gene in ovarian cancers and an insertion/deletion polymorphism in its promoter region. *Cancer Res* 59: 4225–4227

34 Zhang BP, Ye S, Herrmann SM, Eriksson P, de Maat M, Evans A, Arveiler D, Luc G, Cambien F, Hamsten A et al (1999) Functional polymorphism in the regulatory region

of gelatinase B gene in relation to severity of coronary atherosclerosis. *Circulation* 99: 1788–1794

35 Shimajiri S, Arima N, Tanimoto A, Murata Y, Hamada T, Wang KY, Sasaguri Y (1999) Shortened microsatellite d(CA)21 sequence down-regulates promoter activity of matrix metalloproteinase-9 gene. *FEBS Letters* 455: 70–74

36 Okinaga S, Takahashi K, Takeda K, Yoshizawa M, Fujita H, Sasaki H, Shibahara S (1996) Regulation of human heme oxygenase-1 gene expression under thermal stress. *Blood* 87: 5074–5084

37 Yamada N, Yamaya M, Okinaga S, Nakayama K, Sekizawa K, Shibahara S, Sasaki H (2000) Microsatellite polymorphism in the heme oxygenase-1 gene promoter is associated with susceptibility to emphysema. *Am J Hum Genet* 66: 187–195

38 He JQ, Ruan J, Connett J, Anthonisen N, Paré P, Sandford A (2002) Antioxidant gene polymorphisms and susceptibility to a rapid decline in lung function in smokers. *Am J Respir Crit Care Med* 166: 323–328

39 Smith CA, Harrison DJ (1997) Association between polymorphism in gene for microsomal epoxide hydrolase and susceptibility to emphysema. *Lancet* 350: 630–633

40 Yim JJ, Park GY, Lee CT, Kim YW, Han SK, Shim YS, Yoo CG (2000) Genetic susceptibility to chronic obstructive pulmonary disease in Koreans: combined analysis of polymorphic genotypes for microsomal epoxide hydrolase and glutathione S-transferase M1 and T1. *Thorax* 55: 121–125

41 Baranova H, Perriot J, Albuisson E, Ivaschenko T, Baranov VS, Hemery B, Mouraire P, Riol N, Malet P (1997) Peculiarities of the GSTM1 0/0 genotype in French heavy smokers with various types of chronic bronchitis. *Hum Genet* 99: 822–826

42 Harrison DJ, Cantlay AM, Rae F, Lamb D, Smith CA (1997) Frequency of glutathione S-transferase M1 deletion in smokers with emphysema and lung cancer. *Hum Exp Toxicol* 16: 356–360

43 Gilliland FD, Gauderman WJ, Vora H, Rappaport E, Dubeau L (2002) Effects of glutathione-s-transferase m1, t1, and p1 on childhood lung function growth. *Am J Respir Crit Care Med* 166: 710–716

44 Ishii T, Matsuse T, Teramoto S, Matsui H, Miyao M, Hosoi T, Takahashi H, Fukuchi Y, Ouchi Y (1999) Glutathione S-transferase P1 (GSTP1) polymorphism in patients with chronic obstructive pulmonary disease. *Thorax* 54: 693–696

45 Yim JJ, Yoo CG, Lee CT, Kim YW, Han SK, Shim YS (2002) Lack of association between glutathione S-transferase P1 polymorphism and COPD in Koreans. *Lung* 180: 119–125

46 Cantlay AM, Lamb D, Gillooly M, Norrman J, Morrison D, Smith CAD, Harrison DJ (1995) Association between the CYP1A1 gene polymorphism and susceptibility to emphysema and lung cancer. *J Clin Pathol:Mol Pathol* 48: M210–214

47 Huang SL, Su CH, Chang SC (1997) Tumor necrosis factor-α gene polymorphism in chronic bronchitis. *Am J Respir Crit Care Med* 156: 1436–1439

48 Sakao S, Tatsumi K, Igari H, Watanabe R, Shino Y, Shirasawa H, Kuriyama T (2002)

Association of tumor necrosis factor-alpha gene promoter polymorphism with low attenuation areas on high-resolution CT in patients with COPD. *Chest* 122: 416–420

49 Sakao S, Tatsumi K, Igari H, Shino Y, Shirasawa H, Kuriyama T (2001) Association of tumor necrosis factor alpha gene promoter polymorphism with the presence of chronic obstructive pulmonary disease. *Am J Respir Crit Care Med* 163: 420–422

50 Keatings VM, Cave SJ, Henry MJ, Morgan K, O'Connor CM, FitzGerald MX, Kalsheker N (2000) A polymorphism in the tumor necrosis factor-alpha gene promoter region may predispose to a poor prognosis in COPD. *Chest* 118: 971–975

51 Ishii T, Matsuse T, Teramoto S, Matsui H, Miyao M, Hosoi T, Takahashi H, Fukuchi Y, Ouchi Y (2000) Neither IL-1beta, IL-1 receptor antagonist, nor TNF-alpha polymorphisms are associated with susceptibility to COPD. *Respir Med* 94: 847–851

52 Higham MA, Pride NB, Alikhan A, Morrell NW (2000) Tumour necrosis factor-alpha gene promoter polymorphism in chronic obstructive pulmonary disease. *Eur Respir J* 15: 281–284

53 Kucukaycan M, Van Krugten M, Pennings HJ, Huizinga TW, Buurman WA, Dentener MA, Wouters EF (2002) Tumor necrosis factor-alpha +489G/A gene polymorphism is associated with chronic obstructive pulmonary disease. *Respir Res* 3: 29

54 Horne SL, Cockcroft DW, Dosman JA (1990) Possible protective effect against chronic obstructive airways disease by the GC 2 allele. *Hum Hered* 40: 173–176

55 Ishii T, Keicho N, Teramoto S, Azuma A, Kudoh S, Fukuchi Y, Ouchi Y, Matsuse T (2001) Association of Gc-globulin variation with susceptibility to COPD and diffuse panbronchiolitis. *Eur Respir J* 18: 753–757

56 Joos L, McIntyre L, Ruan J, Connett JE, Anthonisen NR, Weir TD, Paré PD, Sandford AJ (2001) Association of IL-1beta and IL-1 receptor antagonist haplotypes with rate of decline in lung function in smokers. *Thorax* 56: 863–866

57 Van Der Pouw Kraan TC, Kucukaycan M, Bakker AM, Baggen JM, Van Der Zee JS, Dentener MA, Wouters EF, Verweij CL (2002) Chronic obstructive pulmonary disease is associated with the -1055 IL-13 promoter polymorphism. *Genes Immun* 3: 436–439

58 He JQ, Connett J, Anthonisen N, Sandford A (2003) Polymorphisms in the IL13, IL13ra1 and IL4RA genes and rate of decline in lung function in smokers. *Am J Respir Cell Mol Biol* 28: 379–385

59 Harries LW, Stubbins MJ, Forman D, Howard GC, Wolf CR (1997) Identification of genetic polymorphisms at the glutathione S-transferase Pi locus and association with susceptibility to bladder, testicular and prostate cancer. *Carcinogenesis* 18: 641–644

60 Cosma G, Crofts F, Taioli E, Toniolo P, Garte S (1993) Relationship between genotype and function of the human CYP1A1 gene. *J Toxicol Environ Health* 40: 309–316

61 Aaron SD, Angel JB, Lunau M, Wright K, Fex C, Le Saux N, Dales RE (2001) Granulocyte inflammatory markers and airway infection during acute exacerbation of chronic obstructive pulmonary disease. *Am J Respir Crit Care Med* 163: 349–355

62 Wilson AG, Symons JA, McDowell TL, McDevitt HO, Duff GW (1997) Effects of a polymorphism in the human tumor necrosis factor alpha promoter on transcriptional activation. *Proc Natl Acad Sci USA* 94: 3195–3199

63 Kamboh MI, Ferrell RE (1986) Ethnic variation in vitamin D-binding protein (GC): a review of isoelectric focusing studies in human populations. *Hum Genet* 72: 281–293

64 Hurme M, Santtila S (1998) IL-1 receptor antagonist (IL-1Ra) plasma levels are co-ordinately regulated by both IL-1Ra and IL-1beta genes. *Eur J Immunol* 28: 2598–2602

65 Zeitlin PL (2000) Future pharmacological treatment of cystic fibrosis. *Respiration* 67: 351–357

66 Gervais R, Lafitte JJ, Dumur V, Kesteloot M, Lalau G, Houdret N, Roussel P (1993) Sweat chloride and ΔF508 mutation in chronic bronchitis or bronchiectasis. *Lancet* 342: 997

67 Dumur V, Lafitte JJ, Gervais R, Debaecker D, Kesteloot M, Lalau G, Roussel P (1990) Abnormal distribution of cystic fibrosis ΔF508 allele in adults with chronic bronchial hypersecretion. *Lancet* 335: 1340

68 Pignatti PF, Bombieri C, Benetazzo M, Casartelli A, Trabetti E, Gile LS, Martinati LC, Boner AL, Luisetti M (1996) CFTR gene variant IVS8-5T in disseminated bronchiectasis. *Am J Hum Genet* 58: 889–892

69 Bombieri C, Benetazzo M, Saccomani A, Belpinati F, Gile LS, Luisetti M, Pignatti PF (1998) Complete mutational screening of the CFTR gene in 120 patients with pulmonary disease. *Hum Genet* 103: 718–722

70 Tzetis M, Efthymiadou A, Strofalis S, Psychou P, Dimakou A, Pouliou E, Doudounakis S, Kanavakis E (2001) CFTR gene mutations – including three novel nucleotide substitutions – and haplotype background in patients with asthma, disseminated bronchiectasis and chronic obstructive pulmonary disease. *Hum Genet* 108: 216–221

71 Cuppens H, Lin W, Jaspers M, Costes B, Teng H, Vankeerberghen A, Jorissen M, Droogmans G, Reynaert I, Goossens M et al (1998) Polyvariant mutant cystic fibrosis transmembrane conductance regulator genes. The polymorphic (Tg)m locus explains the partial penetrance of the T5 polymorphism as a disease mutation. *J Clin Invest* 101: 487–496

72 Matsushita I, Hasegawa K, Nakata K, Yasuda K, Tokunaga K, Keicho N (2002) Genetic variants of human beta-defensin-1 and chronic obstructive pulmonary disease. *Biochem Biophys Res Commun* 291: 17–22

73 Guo X, Lin HM, Lin Z, Montano M, Sansores R, Wang G, DiAngelo S, Pardo A, Selman M, Floros J (2001) Surfactant protein gene A, B, and D marker alleles in chronic obstructive pulmonary disease of a Mexican population. *Eur Respir J* 18: 482–490

74 Kirkbride HJ, Bolscher JG, Nazmi K, Vinall LE, Nash MW, Moss FM, Mitchell DM, Swallow DM (2001) Genetic polymorphism of MUC7: allele frequencies and association with asthma. *Eur J Hum Genet* 9: 347–354

75 Vinall LE, Fowler JC, Jones AL, Kirkbride HJ, de Bolos C, Laine A, Porchet N, Gum JR, Kim YS, Moss FM et al (2000) Polymorphism of human mucin genes in chest disease: possible significance of MUC2. *Am J Respir Cell Mol Biol* 23: 678–686

76 Daly MJ, Rioux JD, Schaffner SF, Hudson TJ, Lander ES (2001) High-resolution haplotype structure in the human genome. *Nat Genet* 29: 229–232

77 Weiss KM, Clark AG (2002) Linkage disequilibrium and the mapping of complex human traits. *Trends Genet* 18: 19–24

The pathology of COPD

Simonetta Baraldo, Renzo Zuin and Marina Saetta

Department of Clinical and Experimental Medicine, Section of Respiratory Diseases, University of Padova, Via Giustiniani 3, 35128 Padova, Italy

Introduction

Chronic obstructive pulmonary disease (COPD) is a complex condition, that includes different disease entities such as bronchiolitis, emphysema and chronic bronchitis. Although the first mention of emphysema dates back to the early 17th Century, progress in our understanding of the pathology of this disease has been regrettably slow.

The relevance of the chronic inflammatory response to the development of COPD has only recently been appreciated and this notion has now been included in the disease definition [1]. In fact, COPD is defined as a disease state characterised by airflow limitation that is not fully reversible. The airflow limitation is usually both progressive and associated with an abnormal inflammatory response of the lungs to noxious particles or gases, the most common of which is tobacco smoke.

The chronic airflow limitation characteristic of COPD is caused by a mixture of small airway disease (bronchiolitis) and parenchymal destruction (emphysema), the relative contributions of which vary from person to person. In fact, since expiratory flow is the result of a driving pressure that promotes flow (elastic recoil of the lung parenchyma), and of an opposing resistance that contrasts flow (obstruction of the airways), a reduction in flow can occur either by reducing the driving pressure or by increasing the resistance. In smokers, when pathological changes involve the small airways, they will contribute to airflow limitation by narrowing and obliterating the lumen and by actively constricting the airways, therefore increasing the resistance. Conversely, when pathological changes are localised in lung parenchyma, they will contribute to airflow limitation by reducing the elastic recoil of the lung, through parenchymal destruction, as well as by reducing the elastic load applied to the airways through destruction of alveolar attachments, therefore reducing the driving pressure.

In this chapter we will focus on pathological changes involving the small airways and lung parenchyma of smokers with COPD, in an attempt to enlighten the mechanisms through which such pathological changes can contribute to the development and the progression of airflow limitation in smokers.

Pathology of small airways in COPD

Even in the absence of an established airflow obstruction, cigarette smoking can induce an inflammatory reaction involving the entire tracheobronchial tree. In particular, it has been demonstrated that young smokers who experienced sudden death outside the hospital have pathological lesions in their small airways [2]. These early lesions include an inflammatory infiltrate in the airway wall consisting predominantly of mononuclear cells and clusters of macrophages in the respiratory bronchioles. These results support the idea that early structural changes may occur in peripheral airways of smokers before COPD is established, probably representing a non-specific response of airways to injury in general. Although the majority of smokers develop a chronic, non-specific airway inflammation, for unknown reasons only some smokers develop overt airflow limitation. The susceptibility factors for the development of COPD are still poorly understood and may involve both genetic traits (i.e., polymorphisms in pro-inflammatory cytokines and genetic control of the balance of helper and cytotoxic T lymphocytes) and environmental conditions triggering or maintaining the disease (i.e., viral infections and pollutants) [3].

In these "susceptible" smokers the inflammatory response designed to protect the lung from the injury of cigarette smoke progresses unrestrictedly and eventually becomes noxious to the structures it was intended to defend. An abnormal infiltration of CD8 T lymphocytes is a key hallmark of COPD pathology, and has been reported in both large and small airways, in alveolar walls and in pulmonary arteries of smokers with COPD [4–6]. In all these lung compartments, CD8 T lymphocytes not only increase in number, but are also inversely related to the degree of airflow limitation, suggesting that CD8 T lymphocytes may play a role in the development of airflow limitation in smokers.

To better characterize the nature of the inflammatory response present in COPD, the pattern of cytokine profile and chemokine receptor expression has recently been investigated [7]. A current paradigm in immunology is that the nature of an immune response to an antigenic stimulus is largely determined by the cytokines produced by activated T cells. Type-1 T cells express cytokines, such as interferon γ (IFN-γ), crucial in the activation of macrophages and in the response to viral and bacterial infections, whereas type-2 T cells express cytokines, such as interleukin (IL)-4 and IL-5, involved in Ig-E mediated responses and eosinophilia characteristic of allergic diseases [8]. It has recently been shown that the CD8+ T cells infiltrating the peripheral airways in COPD produce IFN-γ and express CXCR3 [7], a chemokine receptor that is known to be preferentially expressed on type 1 cells [9]. Moreover

CXCR3 expression is paralleled by a strong epithelial expression of its ligand CXCL10, suggesting that the CXCR3/CXCL10 axis may be involved in the recruitment of type-1 cells in peripheral airways of smokers with COPD.

Along with the progression of chronic airflow limitation, the inflammatory process is associated to the appearance of structural abnormalities in peripheral airways. These structural changes (also referred to as remodelling) include goblet cell hyperplasia and squamous metaplasia in the epithelium, as well as increased airway smooth muscle and fibrosis in the airway wall [10]. Airway wall fibrosis and increased smooth muscle can profoundly affect the mechanical properties of the small airways, promoting airway narrowing. Moreover, inflammation and remodelling of small airways, by increasing the thickness of the airway wall, may facilitate uncoupling between airways and parenchyma, therefore causing airway closure [11, 12]. The inflammatory cells infiltrating the airway wall could also contribute to the destruction of alveolar attachments, i.e., the alveolar walls directly attached to the bronchiolar wall. Indeed, the hypothesis that airway inflammation may play a role in this destructive process is supported by the observation that destruction of alveolar attachments is correlated with the degree of peripheral airway inflammation in smokers [13].

The role of peripheral airway remodelling in the development of airflow limitation characteristic of COPD is highlighted by the pioneering observation of Hogg and co-workers, who first demonstrated that peripheral airways are the major site of increased resistance in the lungs of smokers [14]. As reviewed in this section, pathological changes in small airways contribute to airflow limitation either by increasing the thickness of the airway wall or by decreasing the interdependence between airways and parenchyma, especially through the destruction of alveolar attachments.

Pathology of lung parenchyma in COPD

As pointed out earlier, small airway disease (bronchiolitis) and parenchymal destruction (emphysema) are both important hallmarks of COPD. In particular, emphysema can contribute to the development of airflow limitation by reducing the elastic recoil of the lung, which decreases the intra-alveolar pressure that drives exhalation. Emphysema is defined anatomically as a condition of the lung characterized by permanent abnormal enlargement of the respiratory airspaces, accompanied by destruction of their walls without obvious fibrosis [15]. However, the presence of fibrosis in emphysema is currently a matter of debate, and this part of the definition should probably be rephrased. In fact, in emphysematous lungs, collagen content appears to be increased and microscopic fibrosis is present [16, 17], possibly as a consequence of an attempt of the surrounding tissue to repair emphysematous lesions [18].

Smokers can develop two main morphological forms of emphysema, that can be distinguished according to the region of the acinus involved. Centriacinar (or centrilobular) emphysema is characterised by focal destruction restricted to respiratory bronchioles and to the central portions of the acinus, surrounded by areas of normal lung parenchyma. In this form of emphysema, pathological lesions occur more frequently in the upper lobes of the lung. Panacinar emphysema is characteristic of patients who develop emphysema relatively early in life, and is usually associated with deficiency of α_1-antitrypsin, which normally protects the respiratory region by forming a highly effective anti-elastase screen. This form of emphysema occurs more frequently in the lower lobes than in the upper ones and involves destruction of the alveolar walls in a fairly uniform manner. The most common type of parenchymal destruction in smokers is centriacinar emphysema, but also the panacinar form can be observed.

Centriacinar and panacinar emphysema have distinct functional properties and peripheral airway involvement. The panacinar form is characterized by a higher compliance, while the centriacinar form by a higher degree of hyperreactivity and airway inflammation. It is conceivable that the inflammatory process present in peripheral airways of smokers could favour centriacinar destruction and the consequent development of airflow obstruction observed in centriacinar emphysema. By contrast, in panacinar emphysema, airflow obstruction seems to be due mainly to loss of elastic recoil and to have little relation to peripheral airway inflammation [19, 20].

The pathogenesis of parenchymal destruction remains enigmatic although a mechanism involving a protease-antiprotease imbalance is widely supported. This hypothesis is based on the observation that activated inflammatory cells release proteases which, overwhelming local antiprotease activity, can destroy lung parenchyma. Smoking promotes an inflammatory reaction characterised by an increase in neutrophils and macrophages, that are potential sources of proteases that can damage lung cells and degrade the interstitium (e.g., elastin, collagen, proteoglycans). The molecular mechanisms responsible for this event are now starting to become clear. In particular, it has recently been shown that macrophages, through the release of matrix metalloprotease (MMP)-12 and TNF-α, can selectively recruit large numbers of neutrophils, which can degrade elastin and collagen fibers [21, 22]. However, since many cigarette smokers do not develop significant lung destruction despite a striking inflammatory process, the protease-antiprotease hypothesis may not fully explain the loss of lung tissue in cigarette smoking-induced emphysema.

In COPD, parenchymal destruction is associated with the presence of an inflammatory process in the alveolar walls [23, 24], which consists predominantly of T lymphocytes, particularly CD8+ T lymphocytes [6, 25]. Traditionally, the major activity of CD8+ cytotoxic T lymphocytes has been considered the rapid resolution of acute viral infections, and viral infections are a frequent occurrence in patients with COPD [4]. The observation that people with frequent respiratory infections in childhood are more prone to develop COPD supports the role of viral infections in this disease [26]. It is conceivable that, in response to repeated viral infections, an excessive recruitment of CD8+ T lymphocytes may occur and damage the lung in susceptible smokers, possibly through the release of TNF-α and perphorins [27]. On the other hand, it is also possible that CD8+ T lymphocytes are able to damage the lung even in the absence of a stimulus such as viral infection, as shown by Enelow and co-workers [28], who clearly demonstrated that recognition of a lung "autoantigen" by T cytotoxic cell may directly produce a marked lung injury. Taking into account these findings, it can be hypothesised that the CD8+ cytotoxic T cell accumulation observed in COPD could be a response to an "autoantigenic" stimulus induced by cigarette smoking, possibly during the process of breakdown of matrix components [25, 29].

The observation that CD8+ T cells are increased in lung parenchyma of smokers with COPD and are correlated with the degree of airflow obstruction is intriguing and supports the notion that tissue injury may be dependent on T cell activity [4, 29]. One of the most important consequences of the effects of cytotoxic CD8+ T lymphocytes is the apoptosis of target cells, and it would not be surprising if apoptosis plays a role in the destruction of lung tissue in patients with emphysema. Majo and co-workers [25] have reported that, in smokers with emphysema, both the degree of apoptosis and the number of CD8+ T cells in the alveolar walls increase in parallel with the amount of cigarette smoke inhaled. Moreover Kasahara and co-workers demonstrated that the destruction of lung tissue in emphysema may involve accelerated apoptosis of endothelial and epithelial cells through a mechanism dependent on vascular endothelial growth factor (VEGF) [30]. It can therefore be hypothesised that cytotoxic CD8+ T lymphocytes may participate, along with neutrophils and macrophages, in the destruction of the lungs by inducing apoptosis of structural cells.

The inflammatory process initiated by cigarette smoking advances unrestrictedly as airflow limitation progressively worsens, reaching its climax in end stage disease, where a dramatic amplification of the inflammation has been reported. Retamales and colleagues were the first to examine lung pathology of living patients with severe emphysema [31], reporting an increase in the intensity of the inflammatory response in these subjects. Essentially, in smokers with severe emphysema, all inflammatory cells types (i.e., macrophages, neutrophils, CD4+ and CD8+ T lymphocytes) were increased in the alveolar walls and alveolar spaces [31]. In a recent study we extended these findings by demonstrating that, when the disease progresses towards a very severe stage, there is an amplification of the inflammatory response even in the peripheral airways. This enhanced airway inflammatory process is correlated with the degrees of airflow limitation, lung hyperinflation, CO diffusion impairment and radiologic emphysema, suggesting a role for this inflammatory response in the clinical progression of the disease [32].

Reversibility of pathological lesions

In view of the worldwide increasing prevalence of COPD it is crucial and urgent to identify strategies that may halt the progression of the disease. The question whether interventional approaches (i.e., smoking cessation, use of anti-inflammatory drugs) may reverse the pathological lesions present in COPD is still debated, and only a few studies performed a direct assessment of airway inflammation after quitting smoking or after anti-inflammatory therapy.

Although smoking cessation is the only intervention proven to modify the progressive decline of lung function, quitting smoking does not appear to result in resolution of the airway inflammatory response [33, 34]. This suggests that there are perpetuating mechanisms that maintain the chronic inflammatory process characteristic of COPD, once it has become established [35].

The role of pharmacological interventions in modifying the natural history of COPD is still debated. Long-acting β-agonists improve lung function and health status in COPD patients and reduce exacerbations [36, 37]. Recent studies have shown that, in COPD patients, inhaled corticosteroids have no influence on long-term decline in lung function [38], although they can reduce the incidence of exacerbations [39]. A possible explanation for the effectiveness of corticosteroids in the exacerbations of COPD is the finding that the pattern of bronchial inflammation changes during an exacerbation, showing a prominent airway eosinophilia [40, 41]. The idea that eosinophilic inflammation is a marker for responsiveness to corticosteroids is supported by the observation that airway eosinophilia is present in a subgroup of patients with COPD who improve their pulmonary function in response to a short course of steroids [42, 43]. Moreover, a recent report showed that corticosteroids reduced the number of mast cells in biopsies of COPD patients, and this reduction was paralleled by a decrease in the exacerbation incidence [44]. These findings suggest the presence of a subgroup of patients with COPD characterised by "asthmatic features", i.e., eosinophil and mast cell infiltration, who are responsive to corticosteroids. It is therefore conceivable that corticosteroids may be effective on the pattern of inflammation characteristic of asthma, driven by CD4 T lymphocytes, eosinophils, and mast cells, but not on that characteristic of COPD, driven by CD8 T lymphocytes, macrophages and neutrophils. Only when COPD subjects show "asthmatic features", corticosteroid therapy may be worth considering.

It would seem reasonable to think that, while lesions such as airway inflammation could be potentially reversible, lesions such as parenchymal destruction and fibrosis may be not. In recent years a renewed interest has emerged towards strategies directed at the restoration of the lost alveolar surface area in emphysema. Retinoids have received considerable attention as alveolar morphogens and as potential therapeutic agents, especially because retinoic acid was able to reverse the emphysematous lesions induced by intratracheal administration of elastase in rats

[45]. Nevertheless, given the different mechanisms of lung development in rats and humans, more studies are needed to establish the clinical utility of retinoids in the treatment of human emphysema.

References

1 Pauwels RA, Buist AS, Calverley PM, Jenkins CR, Hurd SS (2001) Global strategy for the diagnosis, management, and prevention of chronic obstructive pulmonary disease. NHLBI/WHO Global Initiative for Chronic Obstructive Lung Disease (GOLD) Workshop summary. *Am J Respir Crit Care Med* 163: 1256–1276
2 Niewoehner DE, Klienerman J, Rice D (1974) Pathological changes in the peripheral airways of young cigarette smokers. *N Engl J Med* 291: 755–758
3 Barnes PJ (2000) Chronic obstructive pulmonary disease. *N Engl J Med* 343: 269–280
4 O'Shaughnessy TC, Ansari TW, Barnes NC, Jeffery PK (1997) Inflammation in bronchial biopsies of subjects with chronic bronchitis: inverse relationship of CD8+ T lymphocytes with FEV1. *Am J Respir Crit Care Med* 155: 852–857
5 Saetta M, Di Stefano A, Turato G, Facchini FM, Corbino L, Mapp CE, Maestrelli P, Ciaccia A, Fabbri LM (1998) CD8+ T-lymphocytes in peripheral airways of smokers with chronic obstructive pulmonary disease. *Am J Respir Crit Care Med* 157: 822–826
6 Saetta M, Baraldo S, Corbino L, Turato G, Braccioni F, Rea F, Cavallesco G, Tropeano G, Mapp CE, Maestrelli P et al (1999) CD8+ve cells in the lungs of smokers with chronic obstructive pulmonary disease. *Am J Respir Crit Care Med* 160: 711–717
7 Saetta M, Mariani M, Panina-Bordignon P, Turato G, Buonsanti C, Baraldo S, Bellettato CM, Papi A, Corbetta L, Zuin R et al (2002) Increased expression of the chemochine receptor CXCR3 and its ligand CXCL-10 in peripheral airways of smokers with chronic obstructive pulmonary disease. *Am J Respir Crit Care Med* 165: 1404–1409
8 Romagnani S (1997) The Th1/Th2 paradigm. *Immunol Today* 18: 263–266
9 Sallusto F, Lenig D, Mackay CR, Lanzavecchia A (1998) Flexible programs of chemokine receptor expression on human polarized T-helper 1 and 2 lymphocytes. *J Exp Med* 187: 875–883
10 Cosio MG, Ghezzo H, Hogg JC, Corbin R, Loveland M, Dosman J, Macklem PT (1977) The relationships between structural changes in small airways and pulmonary-function tests. *N Engl J Med* 298: 1277–1281
11 Lambert RK, Wiggs BR, Kuwano K, Hogg JC, Parè PD (1993) Functional significance of increased airway smooth muscle in asthma and COPD. *J Appl Physiol* 74: 2771–2781
12 Macklem PT (1998) The physiology of small airways. *Am J Respir Crit Care Med* 157: S181-183
13 Saetta M, Ghezzo H, Kim WD, King M, Angus GE, Wang NS, Cosio MG (1985) Loss of alveolar attachments in smokers: an early morphometric correlate of lung function impairment. *Am Rev Respir Dis* 132: 894–900

14 Hogg JC, Macklem PT, WM Thurlbeck (1968) Site and nature of airway obstruction in chronic obstructive lung disease. *N Engl J Med* 278: 1355–1360
15 National Heart Lung and Blood Institute (1985) The definition of emphysema: report of a Division of Lung Diseases workshop. *Am Rev Respir Dis* 132: 182–185
16 Lang MR, Fiaux GW, Gillooly M, Stewart JA, Hulmes DJ, Lamb D (1994) Collagen content of alveolar tissue in emphysematous and non-emphysematous lungs. *Thorax* 49: 319–326
17 Vlahovic G, Russell ML, Mercer RR, Crapo JD (1999) Cellular and connective tissue changes in alveolar septal walls in emphysema. *Am J Respir Crit Care Med* 160: 2086–2092
18 Jeffery PK (2001) Remodeling in asthma and chronic obstructive lung disease. *Am J Respir Crit Care Med* 164: S28–S38
19 Kim WD, Eidelman DH, Izquierdo JL, Ghezzo H, Saetta M, Cosio MG (1991) Centrilobular and panlobular emphysema in smokers. Two distinct morphologic and functional entities. *Am Rev Respir Dis* 144: 1385–1390
20 Saetta M, Kim WD, Izquierdo JL, Ghezzo H, Cosio MG (1994) Extent of centriacinar and panacinar emphysema in smokers' lungs: pathological and mechanical implications. *Eur Respir J* 7: 664–671
21 Snider GL (2003) Understanding inflammation in chronic obstructive pulmonary disease: the process begins. *Am J Respir Crit Care Med* 167: 1045–1046
22 Churg A, Wang RD, Tai H, Wang X, Xie C, Dai J, Shapiro SD, Wright JL (2003) Macrophage metalloelastase mediates acute cigarette smoke-induced inflammation *via* tumor necrosis factor-alpha release. *Am J Respir Crit Care Med* 167: 1083–1089
23 Eidelman D, Saetta M, Ghezzo H, Wang NS, Hoidal JR, King M, Cosio MG (1990) Cellularity of the alveolar walls in smokers and its relation to alveolar destruction. Functional implications. *Am Rev Respir Dis* 141: 1547–1552
24 Finkelstein R, Fraser RS, Ghezzo H, Cosio MG (1995) Alveolar inflammation and its relation to emphysema in smokers. *Am J Respir Crit Care Med.* 152: 1666–1672
25 Majo J, Ghezzo H, Cosio MG (2001) Lymphocyte population and apoptosis in the lungs of smokers and their relationship with emphysema. *Eur Respir J* 17: 946–953
26 Paoletti P, Prediletto R, Carrozzi L, Viegi G, Di Pede F, Carmignani G, Mammini U, Giuntini C, Lebowitz MD (1989) Effects of childhood and adolescence-adulthood respiratory infections in a general population. *Eur Respir J* 2: 428–436
27 Liu AN, Mohammed AZ, Rice WR, Fiedeldey DT, Liebermann JS, Whitsett JA, Braciale TJ, Enelow RI (1999) Perforin-independent CD8(+) T-cell-mediated cytotoxicity of alveolar epithelial cells is preferentially mediated by tumor necrosis factor-alpha: relative insensitivity to Fas ligand. *Am J Respir Cell Mol Biol* 20: 849–858
28 Enelow RI, Mohammed AZ, Stoler MH, Liu AN, Young JS, Lou YH, Braciale TJ (1998) Structural and functional consequences of alveolar cell recognition by CD8(+) T lymphocytes in experimental lung disease. *J Clin Invest* 102: 1653–1661
29 Cosio MG, Majo J, Cosio MG (2002) Inflammation of the airways and lung parenchyma in COPD. Role of T cells. *Chest* 121: 160S–165S

30 Kasahara Y, Tuder RM, Taraseviciene-Stewart L, Lecras TD, Abman S, Hirh PK, Waltenberger J, Voelkel NF (2000) Inhibition of VEGF receptors causes lung cell apoptosis and emphysema. *J Clin Invest* 106: 1311–1319

31 Retamales I, Elliott WM, Meshi B, Coxson HO, Paré PD, Sciurba FC, Rogers RM, Hayashi S, Hogg JC (2001) Amplification of inflammation in emphysema and its association with latent adenoviral infection. *Am J Respir Crit Care Med* 164: 469–473

32 Turato G, Zuin R, Miniati M, Baraldo S, Rea F, Beghe B, Monti S, Formichi B, Boschetto P, Harari S et al (2002) Airway inflammation in severe chronic obstructive pulmonary disease: relationship with lung function and radiologic emphysema. *Am J Respir Crit Care Med* 166: 105–110

33 Turato G, Di Stefano A, Maestrelli P, Mapp CE, Ruggieri MP, Roggeri A, Fabbri LM, Saetta M (1995) Effect of smoking cessation on airway inflammation in chronic bronchitis. *Am J Respir Crit Care Med* 152: 1262–1267

34 Rutgers SR, Postma DS, ten Hacken NHT, Kauffman HF, van der Mark TW, Koeter GH, Timens W (2000) Ongoing inflammation in patients with COPD who do not currently smoke. *Thorax* 55: 12–18

35 Saetta M, Turato G, Maestrelli P, Mapp CE, Fabbri LM (2001) Cellular and structural bases of chronic obstructive pulmonary disease. *Am J Respir Crit Care Med* 163(6): 1304–1309

36 Boyd G, Morice AH, Pounsford JC, Siebert M, Peslis N, Crawford C (1997) An evaluation of salmeterol in the treatment of chronic obstructive pulmonary disease (COPD). *Eur Respir J* 10(4): 815–821

37 Rennard SI, Anderson W, ZuWallack R, Broughton J, Bailey W, Friedman M, Wisniewski M, Rickard K (2001) Use of a long-acting inhaled beta2-adrenergic agonist, salmeterol xinafoate, in patients with chronic obstructive pulmonary disease. *Am J Respir Crit Care Med* 163: 1087–1092

38 Vestbo J, Sorensen T, Lange P, Brix A, Torre P, Viskum K (1999) Long-term treatment with inhaled budesonide in mild and moderate chronic obstructive pulmonary disease: a randomised controlled trial. *Lancet* 353: 1819–1823

39 Burge PS, Calverley PM, Jones PW, Spencer S, Anderson JA, Maslen TK (2000) Randomised, double blind, placebo controlled study of fluticasone proprionate in subjects with moderate to severe chronic obstructive pulmonary disease: the ISOLDE trial. *BMJ* 320: 1297–1303

40 Saetta M, Di Stefano A, Maestrelli P, Turato G, Ruggieri MP, Roggeri A, Calcagni P, Mapp CE, Ciaccia A, Fabbri LM (1994) Airway eosinophilia in chronic bronchitis during exacerbations. *Am J Respir Crit Care Med* 150(6 Pt 1): 1646–1652

41 Saetta M, Di Stefano A, Maestrelli P, Turato G, Mapp CE, Pieno M, Zanguochi G, Del Prete G, Fabbri LM (1996) Airway eosinophilia and expression of interleukin-5 protein in asthma and in exacerbations of chronic bronchitis. *Clin Exp Allergy* 26: 766–774

42 Pizzichini E, Pizzichini MMM, Gibson P, Parameswaran K, Gleich GJ, Berman L, Dolovich J, Heargrave FE (1998) Sputum eosinophilia predicts benefit from prednisone

in smokers with chronic obstructive bronchitis. *Am J Respir Crit Care Med* 158: 1511–1517

43 Chanez P, Vignola AM, O'Shaughnessy T, Enander I, Li D, Jeffery PK, Bousquet J (1997) Corticosteroid reversibility in COPD is related to features of asthma. *Am J Respir Crit Care Med* 155: 1529–1534

44 Hattotuwa KL, Gizycki MJ, Ansari TW, Jeffery PK, NC Barnes (2002) The effects of inhaled fluticasone on airway inflammation in chronic obstructive pulmonary disease. A double-blind, placebo-controlled biopsy study. *Am J Respir Crit Care Med* 165: 1592–1596

45 Massaro GD, Massaro D (1997) Retinoic acid treatment abrogates elastase-induced pulmonary emphysema in rats. *Nature Med* 3: 675–703

Lung function in COPD

Frances Gilchrist[1], Onn Min Kon[1] and Michael I. Polkey[2]

[1]Department of Respiratory Medicine, St Mary's Hospital, Praed Street, London W2 1NY, UK;
[2]Royal Brompton Hospital & National Heart & Lung Institute, Fulham Road, London SW3 6NP, UK

Introduction

Chronic obstructive pulmonary disease (COPD) is characterised by gradual progression of respiratory disability from symptoms that are merely irritating, such as a morning cough, through to breathlessness on minimal exertion, ventilatory failure and death. Physiology is relevant at both extremes of this spectrum though the purpose changes. Specifically for patients in an early stage of disease physiological techniques may be used to diagnose the condition. Early diagnosis allows rational focussing of smoking cessation therapies as well as, in the future, the possible prescription of disease modifying agents. For patients with advanced disease, agents that modify the course of the disease are likely to have limited value even if they are effective in patients with early disease; however it is in these patients that an understanding of physiological mechanisms may allow the prescription of novel non-pharmacologic therapies such as lung volume reduction surgery.

Physiological techniques for diagnosis and assessment in COPD

Forced expiratory volume in 1 second (FEV$_1$)

Spirometry is simple and the most useful technique for diagnosing and monitoring the progression of COPD [1]. The subject exhales from total lung capacity and volume expired with time is recorded [2]. Analysis of the expired volume *versus* time curve allows the FEV$_1$ (forced expiratory volume in 1 second), the forced vital capacity (FVC) and the FEV$_1$/FVC ratio to be derived [3]. Figure 1 illustrates a normal expired volume *versus* time curve and those found in mild COPD (panel b) and more serious disease (panel c). COPD is defined as: "a chronic, slowly progressive disorder characterised by airway obstruction which does not change markedly over several months. The impairment of lung function is largely fixed but is partially reversible by bronchodilator or other therapy" [1].

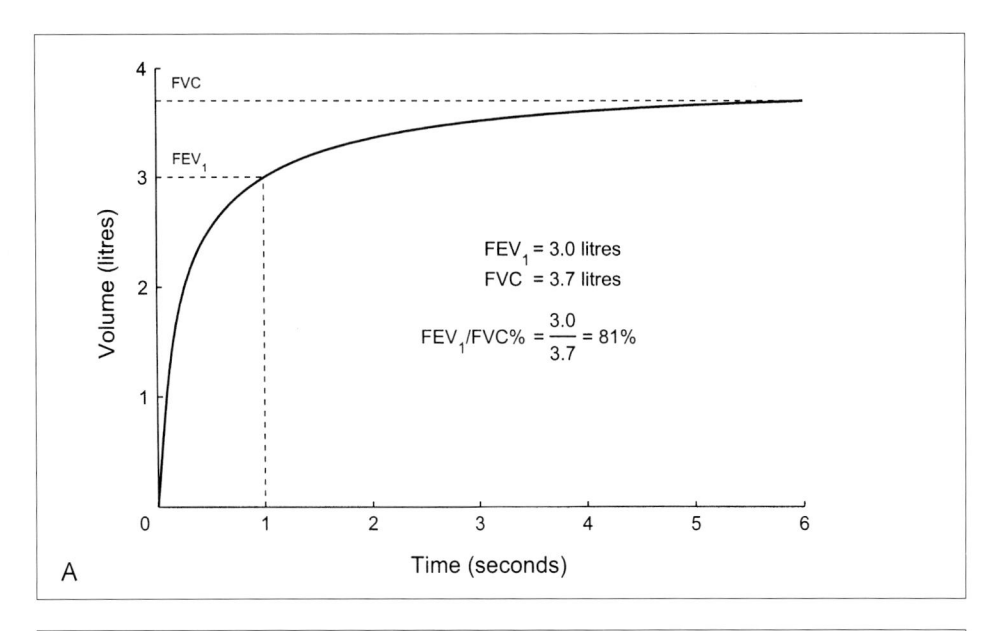

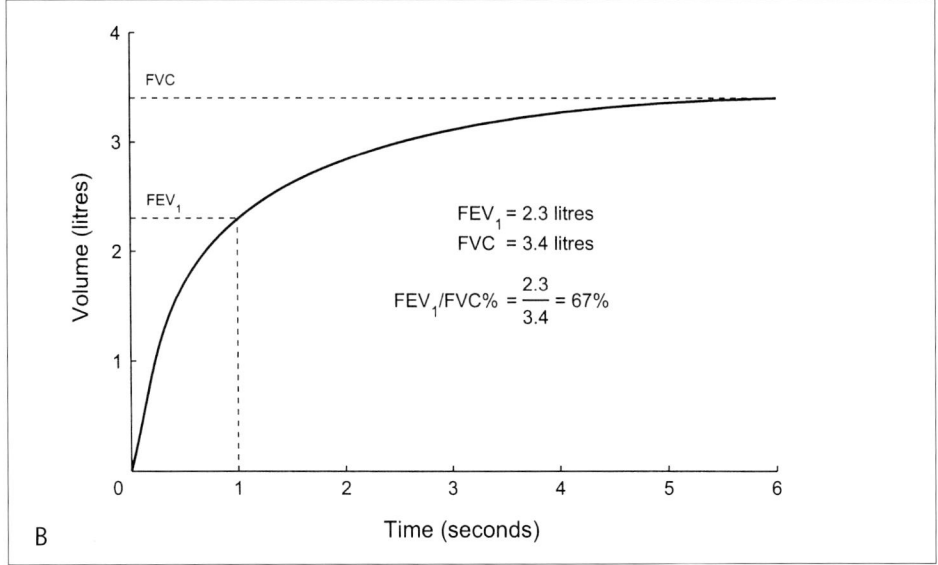

Figure 1
A. Expiratory volume versus time curve obtained by spirometry in a normal subject.
B. Expiratory volume versus time curve in a subject with mild COPD. The FEV_1/FVC ratio has fallen to 67%.
C. The expiratory volume versus time curve in a subject with severe COPD. The FEV_1/FVC ratio is now only 22% indicating very marked airways obstruction.

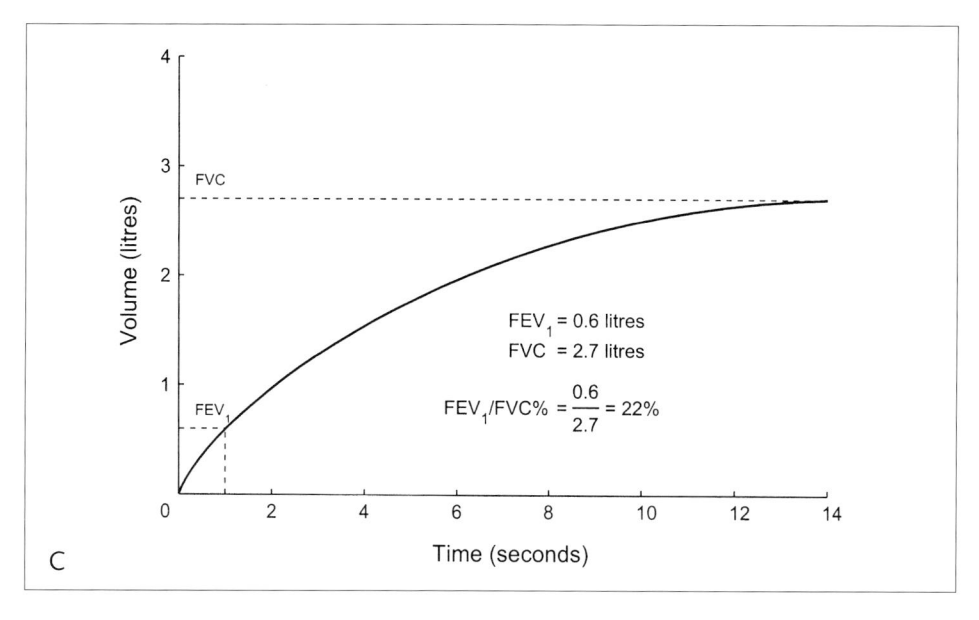

[Figure: Volume-time spirometry curve showing FVC = 2.7 litres, FEV₁ = 0.6 litres, FEV₁/FVC% = 0.6/2.7 = 22%]

Figure 1 (continued)

An FEV_1/FVC ratio of less than 70% is diagnostic of airways obstruction. Reversibility can be assessed by giving a bronchodilator such as 400 micrograms of a $β_2$-agonist or 80 µg of an anticholinergic [4, 5]. An FEV_1 of less than 80% predicted at 30 minutes post-bronchodilator confirms that airways obstruction is not fully reversible while an increase in FEV_1 of 200 mls or 12% above pre-bronchodilator level is considered to show significant reversibility [6, 7].

The FEV_1 is a highly reproducible test and is used to monitor disease progression. It can also be used to assess disease severity as shown in Table 1 [1].

This classification is of use in making treatment decisions. FEV_1 is favoured as a monitoring tool to assess progression of the condition as, although the FEV_1/FVC ratio is useful in diagnosis, the FEV_1 is more reproducible than FVC. Some patients with obstructive lung disease also have a reduced VC because of co-existent disease (most commonly obesity) and a FEV_1/VC ratio > 70% does not preclude COPD.

Peak expiratory flow

The peak expiratory flow is the maximum flow rate obtained during a forced expiration starting from total lung capacity. It is easily measured on a readily available and portable meter. However, it is effort-dependent which affects its reproducibility and in COPD often underestimates the degree of airflow limitation [8]. This is

Table 1 - GOLD classification of COPD

Disease stage	Characteristics
0: At risk	Normal spirometry
	Chronic symptoms (cough, sputum production)
1: Mild COPD	$FEV_1/FVC < 70\%$
	$FEV_1 > 80\%$ predicted
	With or without chronic symptoms (cough, sputum production)
II: Moderate COPD	$FEV_1/FVC < 70\%$
	$30\% < FEV_1 < 80\%$ predicted
	(IIA: $50\% < FEV_1 < 80\%$ predicted)
	(IIB: $30\% < FEV_1 < 50\%$ predicted)
	With or without chronic symptoms (cough, sputum production)
III: Severe COPD	$FEV_1/FVC < 70\%$
	$FEV_1 < 30\%$ predicted or $FEV_1 < 50\%$ predicted plus respiratory failure or clinical signs of right heart failure

explained by the airflow *versus* volume curve for COPD (Fig. 2). The peak expiratory flow occurs during the first 100 ms of expiration. In severe COPD this early flow is comparatively preserved but then pressure-dependent airways collapse leading to a rapid falling off in flow. This is illustrated in Figure 2, which compares the smooth normal flow/volume loop with that for COPD. The COPD curve has two distinct phases: during the early stages of expiration flow is slightly reduced compared with normal; however, later in expiration, the intrapleural pressure exceeds that holding the airways open and they collapse. This is due to a reduction in the static recoil pressure of the lung resulting from parenchymal destruction [9]. This gives the flow/volume loop of COPD its characteristic "steepled" shape and causes a dramatic reduction in flow during the later stages of expiration.

Lung volume assessments

Inert gas techniques

There are two methods both making use of inert gases, which are commonly used to assess lung volumes. These allow measurement of a number of different indices:

- The total lung capacity (TLC) which is the volume of gas within the lungs at maximum inspiration occurring when the outward pull generated by the inspiratory muscles balances the combined inward recoil of the lungs and chest wall.

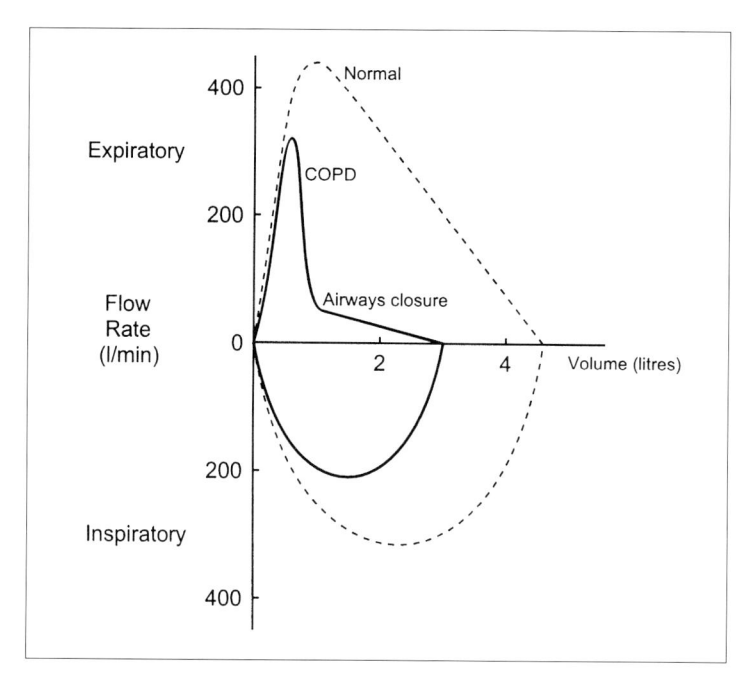

Figure 2
The flow volume loops of a normal subject (dotted line) and a subject with COPD (continuous line) showing the characteristic steepled shape caused by pressure dependent airways collapse.

- The functional residual volume (FRC) or volume of air remaining in the lungs after a quiet expiration. This represents the mechanically neutral position of the chest with complete relaxation of the respiratory muscles. At this point, theoretically the inward recoil of the lung is balanced by the outward pressure exerted by the chest wall. In fact patients with advanced COPD may never reach true FRC because exhalation is interrupted by the subsequent inspiration [10] and in such patients the correct physiological term is end-expiratory lung volume. The residual volume (RV) is the volume of air remaining after a maximal expiration or the point at which the inward force generated by expiratory muscles and lungs balances the outwards recoil of the chest wall.

The relationship between these volumes is illustrated by Figure 3. In COPD, particularly if there is significant emphysema, lung volumes are often abnormal. COPD often leads to an increase in TLC, FRC and RV and the RV/TLC ratio which is less than 30% in young adults with normal lungs but can be greater than 40% due to increased lung compliance and reduced lung recoil pressure [11]. There is little evi-

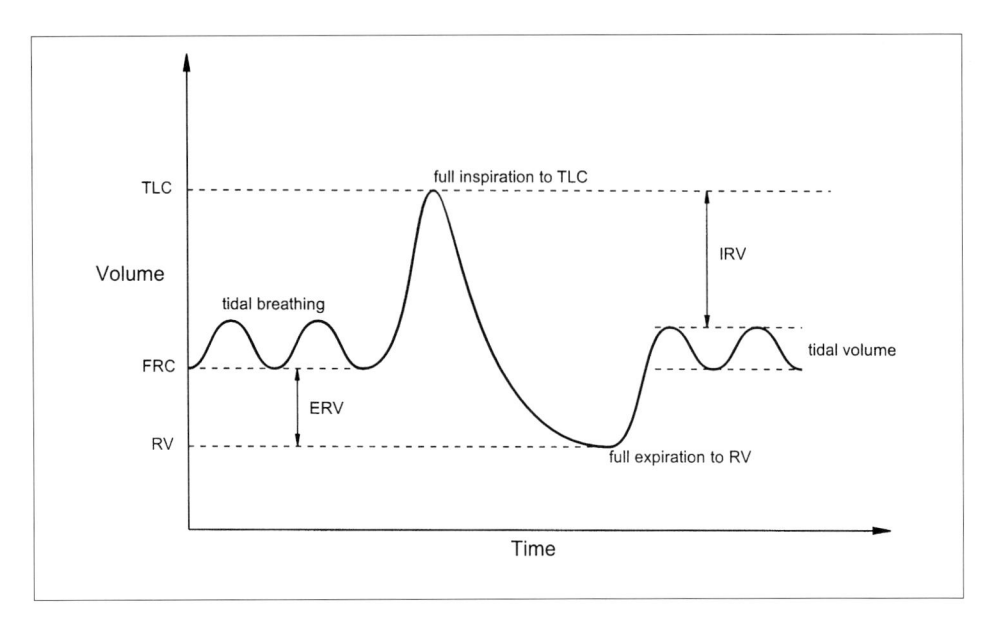

Figure 3
Spirometric trace illustrating the relationships between the commonly measured static lung volumes. RV, residual volume; FRC, functional residual volume; ERV, expiratory reserve volume; TLC, total lung capacity; IRV, inspiratory reserve volume.

dence that routine assessment of lung volumes in COPD is useful, however, it is does allow assessment of hyperinflation and is important if lung volume reduction surgery is considered [12].

Lung volumes can be measured using either the nitrogen washout or helium dilution techniques [2]. Both rely on the fact that at constant temperature the product of the concentration and volume of a gas is also constant. The nitrogen washout method uses the nitrogen already present in air as the inert indicator gas. The subject breathes oxygen for 7–10 minutes and the exhaled gases are collected. The functional residual capacity (FRC) is computed using the formula:

$$C1V1 = C2V2$$

where C1 is the initial nitrogen content of exhaled gas which is assumed to be that of air, V1 is the FRC, C2 the final nitrogen content of the exhaled gas and V2 the total volume of exhaled gas. Alternatively the nitrogen content of the exhaled gas can be calculated second-by-second using a continuous measuring device in the mouthpiece. Calculating FRC involves rather more complicated calculations using this method but it has the advantage of giving some indication of non-uniform ventilation of alveoli [2].

The helium dilution technique uses added helium as the indicator gas. The air within the lungs is allowed to equilibrate with air containing a known small concentration of helium. Practically the patient is asked to breath through a closed spirometer of known volume containing a known concentration of helium. The end point is reached when the helium concentration falls by < 0.02% in 30 seconds. In an individual with normal lungs this takes 5–10 minutes. However, in COPD equilibration occurs far more slowly. This means it can take a considerable time to reach this point. This is a disadvantage of this method. The final concentration of the helium in the spirometer allows calculation of the volume of the lungs at the time the subject is connected to the circuit. Usually this is chosen to be FRC but a correction factor for the volume of the apparatus and anatomical dead space has to be included. At the end of the procedure the subject is asked to inspire as far as possible. The value obtained is added to the FRC measurement to give TLC [2]. In COPD this method tends to underestimate lung volumes because poorly ventilated areas of the lung due to airway distortion, bullae and poorly communicating alveoli equilibrate much more slowly with the inhaled gas than normally ventilated lung. However the difference between alveolar volume so determined and the plethysmographic TLC (*vide infra*) may give insight into the magnitude of gas trapping.

Lung volumes – plethysmographic

A second approach to measuring lung volumes is to use plethysmography. This technique is based on Boyles Law that at constant temperature, the product of the pressure of the gas and its volume is constant. The subject sits in a sealed box and makes gentle inspiratory efforts against a closed shutter at the mouthpiece. The shutter ensures no movement of gas out of the lungs. A reduction of gas pressure in the lungs increases their volume which in turn increases the gas pressure in the box. Thus box pressure is proportional to lung volume [2]. This method allows determination of functional residual volume (FRC), and other lung volumes can be derived from this by a range of inspiratory and expiratory manoeuvres. This method is more accurate than the inert gas techniques for assessing lung volumes in COPD as it does not rely on ventilation and also has the advantage of being rapidly performed. However, it requires very expensive equipment which needs highly skilled operators and also requires a high degree of patient co-operation. Also, this method assumes pressure changes measured at the mouth and changes in alveolar pressure are identical. This assumption is less likely to be true in those with COPD and the difference is greater with increasing severity of airways obstruction. This will lead lung volumes to be over-estimated. The box measures all changes in gas pressures occurring in the body. This includes intestinal as well as thoracic gas. Again this will tend to lead lung volumes to be over-estimated. A combination of an inert gas technique together with whole body plethysmography will give the best overall assessment of lung volumes in COPD.

Carbon monoxide gas transfer

The carbon monoxide gas transfer (TLCO or DLCO) is a measure of the ease with which carbon monoxide moves from the alveoli across the alveolar membrane to become attached to haemoglobin. It is used as a method of assessing gas exchange by the lung. It is measured in a number of ways:

- Single breath method [13] – The subject breathes in rapidly from RV to TLC i.e., a vital capacity breath from a reservoir bag containing air together with a small known concentration of carbon monoxide and helium which acts as a reference gas. The breath is then held for approximately ten seconds and the subject exhales rapidly back to RV. The first 750 mls of exhaled gas is discarded as this contains gas derived from the anatomical dead space. The next 500 mls of gas is analysed for the levels of helium and carbon monoxide. The difference in these concentrations compared with the initial concentrations allows the amount of carbon monoxide which has passed across the alveolar membrane and become irreversibly bound to haemoglobin to be calculated. From this, together with the breath hold time and volume of expired gas, the TLCO can be calculated [2].
- Constant exhalation method [14] – As in the single breath method the subject inhales a vital capacity breath from a reservoir bag containing carbon dioxide and a reference gas, usually methane. From maximal inspiration the subject exhales slowly against a fixed resistance. The carbon monoxide in exhaled gas is analysed at a rate of 31 times a second by an infra red sensor.
- Rebreathing method – The subject breathes in and out from a reservoir bag containing known concentrations of carbon dioxide and the reference gas sulphur hexafluoride. Exhaled gases are monitored continuously using a mass spectrometer.

The carbon monoxide transfer factor depends on the conductance of the gas across the alveolar membrane and into the red cell followed by its combination with haemoglobin. This is influenced by the alveolar oxygen concentration as this competes with carbon monoxide for binding to haemoglobin and the haemoglobin concentration. Membrane conductance depends on the surface area available for gas diffusion and on the integrity of the alveolar membrane and its interaction with the capillary. In COPD the transfer factor is frequently reduced. Emphysema leads to significant destruction of pulmonary capillaries and a reduction in the surface area available for gas transfer due to the formation of poorly ventilated areas such as bullae. Smoking means that under normal conditions the subject will have an increased blood concentration of carbon monoxide. This will reduce the diffusion gradient by which carbon monoxide moves in to the blood hence reducing transfer and leading to a lower calculated transfer factor.

Measurement of the transfer factor for carbon dioxide is not a routinely performed test in COPD but it is very useful if lung volume reduction surgery is considered. There is evidence that the pre-operative TLCO predicts how successful lung volume reduction surgery will be. A TLCO of less than 20–30% predicted is associated with increased mortality [12, 15].

Advanced physiologic techniques; findings and implications for therapy

Static pressure volume curves

The normal relationship between pressure and volume for the relaxed respiratory system is sigmoid in shape [16], so that at the middle portion (around FRC) small changes in pressure result in large changes in volume. The pressure volume properties of the respiratory system are the sum of the properties of the lung and chest wall and these may be distinguished *in vivo* by measuring the pressure both at the mouth and in the oesophagus so that the chest wall pressure is the oesophageal pressure and the lung pressure is the transpulmonary pressure (i.e., mouth minus oesophageal pressure). Static compliance (whether lung, chest wall or respiratory system) is the change in volume divided by the change in pressure observed in moving between two points on this curve and by convention is usually measured over the linear portion above FRC, during expiration.

Pressure volume curves are easier to measure in anaesthetised paralysed subjects because one can then be certain that the observed pressures are not influenced by muscle activation. However, with care [17], pressure volume curves can be measured in unanaesthetised naïve patients with COPD. The classic abnormality in emphysema is a reduction in lung recoil pressure and it is this that leads to the rise in lung volumes characteristic of the condition. Compliance has been measured after few interventions in patients with COPD because it is difficult to measure. However, a return towards normal has been shown in emphysematous patients after β_2 agonists and smoking cessation [18] and in those undergoing lung volume reduction surgery [19, 20] (Fig. 4).

Inspiratory capacity/dynamic hyperinflation

When patients with advanced COPD exercise flow limitation occurs [21]. If such a patient cannot expire the previously inhaled tidal volume before the next inspiration then dynamic hyperinflation (i.e., an increase in end-expiratory lung volume) will occur progressively as exercise continues. The magnitude of dynamic hyperinflation

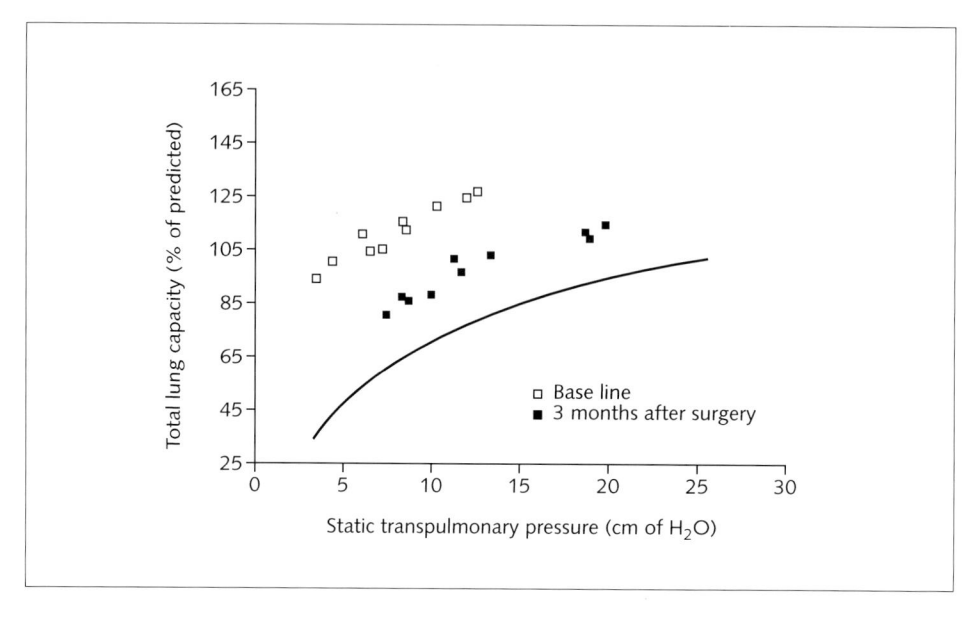

Figure 4
Lung recoil pressure from a 64-years-old woman before and three months after lung volume reduction surgery. The pressure volume curve for the normal lung is shown for comparison. Data from Sciurba et al. [19].

may be of the order of a litre [22]. Presently end-expiratory lung volume (EELV) is difficult to measure in absolute terms, although new techniques, such as opticoelectric plethysmography [23], could facilitate this. However because total lung capacity (TLC) remains constant even during exercise in COPD [24], then reductions in inspiratory capacity (the difference between TLC and EELV) are a good measure of dynamic hyperinflation [25].

Using this technique a reduction in dynamic hyperinflation has been demonstrated to be an important determinant of exercise performance in patients with COPD [26]. Reduction in dynamic hyperinflation may be achieved using oxygen [27], bronchodilators [28] and lung volume reduction surgery [29].

Negative expiratory pressure
The negative expiratory pressure (NEP) technique is an ingenious method to detect flow limitation recently developed by Koulouris et al. [30]. The essence of the technique is that a small sub-atmospheric pressure (~ 5 cm H_2O) applied at the mouth causes a flow increase in patients who are not flow limited (Fig. 5). Using this technique it is possible to more precisely determine flow limitation.

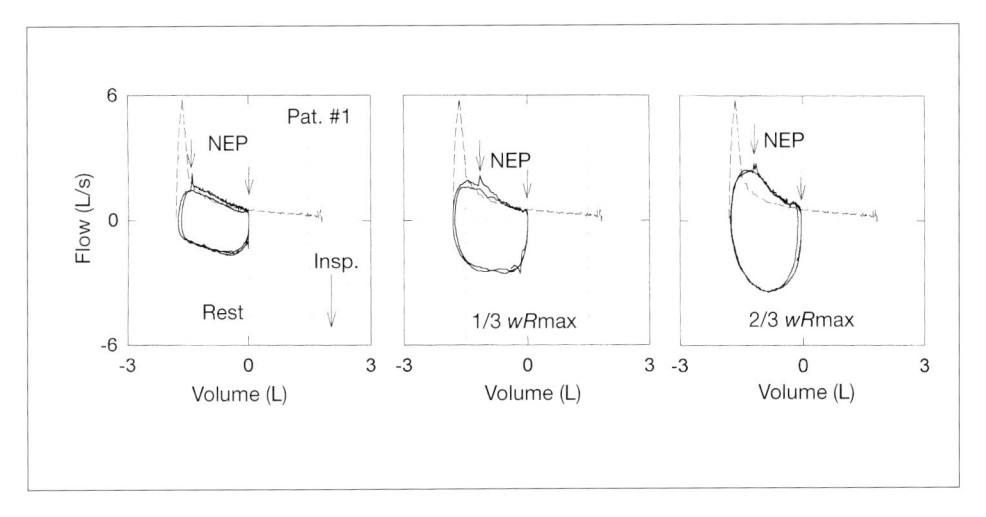

Figure 5
Example of traces obtained using the NEP technique in a patient with COPD. Increased flow following the application of negative pressure (between arrows) is visible at rest but not during exercise (right two panels). Reproduced from Koulouris et al. [52], with permission.

Respiratory muscle function

Because patients with COPD have hyperinflation, shortening of the inspiratory muscles occur. In respiratory muscle, as with other skeletal muscles, shortening is associated with a reduction in tension-generating ability and therefore it is not unexpected that inspiratory muscle pressures are reduced in patients with COPD [31]. Whether the diaphragm (as the most important and most shortened inspiratory muscle) is weaker than would be expected given the observed hyperinflation is more controversial (e.g., [32, 33]); however it is clear that the diaphragm undergoes a switch in fibre type towards more Type I fatigue resistant fibres [34] and that this is probably partially responsible for the observed fatigue resistance in these patients [35, 36].

The effect on diaphragm function of a variety of pharmacological interventions has been evaluated in COPD including β_2 agonists [37–39], methylxanthines [40] and digoxin [41]. No convincing evidence for improvement in inspiratory muscle strength other than those mediated by lung volume change has been observed.

Inspiratory muscle training has also been advocated as a means of improving inspiratory muscle strength in COPD. Although the use of an inspiratory muscle trainer can result in both normal subjects [42, 43] and patients with COPD [44, 45] achieving a more forceful maximal static mouth pressure, there is dispute as to how this effects. Proponents of inspiratory muscle training consider the effect to be mediated by muscle hypertrophy in response to training. However another possibility is either a placebo effect or, alternatively, an alteration in cortical excitability conferring better performance in the maximal static mouth pressure test [46]. This point of view is supported by the lack of improvement in non-volitional measures of diaphragm strength following inspiratory muscle training [43].

Lung volume reduction surgery [47, 48] and lung transplantation [49] confer volume mediated improvements in diaphragm strength. The exact magnitude of functional improvement following LVRS remains to be determined as does the issue of whether diaphragm function could improve patient selection for this procedure.

Sleep hypoxaemia
Significant hypoxaemia during the day (generally < 7.3 kPa) confers a poor prognosis in COPD, which can be partially reversed by continuous supplemental oxygen therapy [50]. Similarly nocturnal hypoventilation in COPD confers a worse prognosis unless treated with supplemental oxygen [51]. Identification of sleep hypoxaemia is relatively straightforward using oximetry alone but these traces frequently fail to differentiate between obstructive events or hypoventilation. This distinction may be clinically important when deciding which form of ventilatory support (continuous positive airways pressure (CPAP), or non-invasive ventilation (NIV)) is most appropriate. One approach to resolving this is polysomnography but this may be expensive and alternative is to measure transcutaneous nocturnal carbon dioxide (CO_2).

Conclusions

New therapies are appearing for patients with COPD at the start and finish of their illness. While creative strategies are required to develop new approaches we believe that assessment of their efficacy can only be achieved through careful physiological measurement which requires fitting the best physiological test to the postulated method of action.

References

1 Pauwels RA, Buist AS, Calverley PM, Jenkins CR, Hurd SS (2001) Global strategy for the diagnosis, management and prevention of chronic obstructive pulmonary disease. NHLBI/WHO global initiative for chronic obstructive lung disease (GOLD) workshop summary. *Am J Respir Crit Care Med* 163: 1256–1276
2 Hughes JMB, Pride NB (1999) *Lung function tests: Physiological principles and clinical applications*. WB Saunders, London

3   American Thoracic Society. 1995 Standardization of Spirometry, 1994 Update. *Am J Respir Crit Care Med* 152: 1107–1136
4   Reis AL (1982) Response to bronchodilators. In JL Clausen (ed): *Pulmonary function testing: guidelines and controversies.* Academic Press, New York, 215–221
5   American Thoracic Society (1991) Lung function testing: selection of reference values and interpretative strategies. *Am Rev Respir Dis* 144: 1202–1218
6   Quanjer PH, Tammeling GJ, Cotes JE, Pedersen OF, Peslin R, Yernault JC (1993) Lung volumes and forced ventilatory flows. Report Working Party Standardization of Lung Function Tests, European Community for Steel and Coal. Official Statement of the European Respiratory Society. *Eur Respir J* (Suppl) 16: 5–40
7   Tweeddale PM, Alexander F, McHardy GJ (1987) Short term variability in FEV1 and bronchodilator responsiveness in patients with obstructive ventilatory defects. *Thorax* 42: 487–490
8   Kelly CA, Gibson GJ (1988) Relation between FEV1 and peak expiratory flow in patients with chronic airflow obstruction. *Thorax* 43: 335–336
9   Mead J, Turner JM, Macklem PT, Little JB (1967) Significance of the relationship between lung recoil and maximum expiratory flow. *J Appl Physiol* 22: 95–108
10 Purro A, Appendini L, Patessio A, Zanaboni S, Gudjonsdottir M, Rossi A, Donner CF (1998) Static intrinsic PEEP in COPD patients during spontaneous breathing. *Am J Respir Crit Care Med* 157(4 Pt 1): 1044–1050
11 Morris MJ, Madgwick RG, Lane DJ (1996) Difference between functional residual capacity and elastic equilbrium volume in patients with chronic obstructive pulmonary disease. *Thorax* 51: 415–419
12 National Emphysema Treatment Triallists (2001) Patients at high risk of death after lung-volume-reduction surgery. *N Engl J Med* 345: 1075–1083
13 Ogilvie CM, Forster RE, Blakemore WS, Morton JW (1957) A standardized breath holding technique for the clinical measurement of the diffusing capacity of the lung for carbon monoxide. *J Clin Invest* 36: 1–17
14 Wilson AF, Hearne J, Brenner M, Alfonso R (1994) Measurement of transfer factor during constant exhalation. *Thorax* 49: 1121–1126
15 Geddes D, Davies M, Koyama H, Hansell D, Pastorino U, Pepper J, Agent P, Cullinan P, MacNeill SJ, Goldstraw P (2000) Effect of lung-volume-reduction surgery in patients with severe emphysema. *N Engl J Med* 343: 239–245
16 Rahn H, Otis AB, Chadwick LE, Fenn WO (1946) The pressure-volume diagram of the thorax and lung. *Am J Physiol* 146: 161–178
17 Moy ML, Loring SH (1998. Compliance. *Semin Resp Crit Care Med* 19: 349–359
18 Michaels R, Sigurdson M, Thurlbeck S, Cherniack R (1979) Elastic recoil of the lung in cigarette smokers: the effect of nebulized bronchodilator and cessation of smoking. *Am Rev Respir Dis* 119: 707–716
19 Sciurba FC, Rogers RM, Keenan RJ, Slivka WA, Gorcsan J, Ferson PF, Holbert JM, Brown ML, Landreneau RJ (1996) Improvement in pulmonary function and elastic

recoil after lung-reduction surgery for diffuse emphysema. *N Engl J Med* 334: 1095–1099

20 Gelb AF, Zamel N, McKenna RJ Jr, Brenner M (1996) Mechanism of short-term improvement in lung function after emphysema resection. *Am J Respir Crit Care Med* 154(4 Pt 1): 945–951

21 Potter WA, Olafsson S, Hyatt RE (1971) Ventilatory mechanics and expirtaory flow limitation during exercise in patients with obstructive lung disease. *J Clin Invest* 50: 910–919

22 Dodd DS Brancatisano T, Engel LA (1984) Chest wall mechanics during exercise in patients with severe chronic airflow obstruction. *Am Rev Respir Dis* 129: 33–38

23 Kenyon CM, Cala SJ, Yan S, Aliverti A, Scano G, Duranti R, Pedotti A, Macklem PT (1997) Rib cage mechanics during quiet breathing and exercise in humans. *J Appl Physiol* 83: 1242–1255

24 Stubbing DG, Pengelly LD, Morse JLC, Jones NL (1980) Pulmonary mechanics during exercise in subjects with chronic airflow obstruction. *J Appl Physiol* 49: 511–515

25 Yan S, Kaminski D, Sliwinski P (1997) Reliability of inspiratory capacity for estimating end-expiratory lung volume changes during exercise in patients with chronic obstructive pulmonary disease. *Am J Respir Crit Care Med* 156: 55–59

26 O'Donnell DE, D'Arsigny C, Fitzpatrick M, Webb KA (2002) Exercise hypercapnia in advanced chronic obstructive pulmonary disease: the role of lung hyperinflation. *Am J Respir Crit Care Med* 166: 663–668

27 O'Donnell DE, D'Arsigny C, Webb KA (2001. Effects of hyperoxia on ventilatory limitation during exercise in advanced chronic obstructive pulmonary disease. *Am J Respir Crit Care Med* 163: 892–898

28 Belman MJ, Botnick WC, Shin JW (1996) Inhaled bronchodilators reduce dynamic hyperinflation during exercise in patients with Chronic Obstructive Pulmonary Disease. *Am J Respir Crit Care Med* 153: 967–975

29 Martinez FJ, de Oca MM, Whyte RI, Stetz J, Gay SE, Celli BR (1997) Lung-volume reduction improves dyspnea, dynamic hyperinflation, and respiratory muscle function. *Am J Respir Crit Care Med* 155: 1984–1990

30 Koulouris NG, Valta P, Lavoie A, Corbeil C, Chasse M, Braidy J, Milic-Emili J (1995) A simple method to detect expiratory flow limitation during spontaneous breathing. *Eur Respir J* 8: 306–313

31 Byrd RB, Hyatt RE (1968) Maximal respiratory pressures in chronic obstructive lung disease. *Am Rev Respir Dis* 98: 848–856

32 Similowski T, Yan S, Gauthier AP, Macklem PT, Bellemare F (1991) Contractile properties of the human diaphragm during chronic hyperinflation. *N Engl J Med* 325: 917–923

33 Polkey MI, Kyroussis D, Hamnegard C-H, Mills GH, Green M, Moxham J (1996) Diaphragm strength in chronic obstructive pulmonary disease. *Am J Respir Crit Care Med* 154: 1310–1317

34 Levine S, Kaiser L, Leferovich J, Tikunov B (1997) Cellular adaptations in the diaphragm in chronic obstructive pulmonary disease. *N Engl J Med* 337: 1799–1806

35 Polkey MI, Kyroussis D, Keilty SEJ, Hamnegard CH, Mills GH, Green M, Moxham J (1995) Exhaustive treadmill exercise does not reduce twitch transdiaphragmatic pressure in patients with COPD. *Am J Respir Crit Care Med* 152: 959–964

36 Polkey MI, Kyroussis D, Hamnegard C-H, Mills GH, Hughes PD, Green M, Moxham J (1997) Diaphragm performance during maximal voluntary ventilation in chronic obstructive pulmonary disease. *Am J Respir Crit Care Med* 155: 642–648

37 Bellamy D, Hutchison DCS (1981)The effects of salbutamol aerosol on lung function in patients with pulmonary emphysema. *Br J Dis Chest* 75: 190–196

38 Stoller JK, Wiedemann HP, Loke J, Snyder P, Virgulto J, Matthay RA (1988) Terbutaline and diaphragm function in chronic obstructive pulmonary disease: a double blind randomized clinical trial. *Br J Dis Chest* 82: 242–250

39 Hatipoglu US, Laghi F, Tobin MJ (1999) Does inhaled albuterol improve diaphragmatic contractility in patients with chronic obstructive pulmonary disease? *Am J Respir Crit Care Med* 160: 1916–1921

40 Lanigan C, Howes TQ, Borzone G, Vianna LG, Moxham J (1993) The effects of beta2-agonists and caffeine on respiratory and limb muscle performance. *Eur Resp J* 6: 1192–1196

41 Liberman D, Brami JL, Bark H, Pilpel D, Heimer D (1991) Effect of digoxin on respiratory muscle performance in patients with COPD. *Respiration* 58: 29–32

42 Leith DE, Bradley M (1976) Ventilatory muscle strength and endurance training. *J Appl Physiol* 41(4): 508–516

43 Hart N, Sylvester K, Ward S, Cramer D, Moxham J, Polkey MI (2001) Evaluation of an inspiratory muscle trainer in healthy humans. *Respir Med* 95: 526–531

44 Sanchez Riera H, Montemayor Rubio T, Ortega Ruiz F, Cejudo Ramos P, Del Castillo Otero D, Elias Hernandez T, Castillo Gomez J (2001) Inspiratory muscle training in patients with COPD: effect on dyspnea, exercise performance, quality of life. *Chest* 120: 748–756

45 Lotters F, van Tol B, Kwakkel G, Gosselink R (2002) Effects of controlled inspiratory muscle training in patients with COPD: a meta-analysis. *Eur Respir J* 20: 570–576

46 Demoule A, Verin E, Derenne J-P, Similowski T (2001) Plasticity of the human motor cortical representation of the diaphragm. *Am J Respir Crit Care Med* 163: A46

47 Laghi F, Jubran A, Topeli A, Fahey P, Garrity ER, Arcidi JM, de Pinto DJ, Edwards LC, Tobin MJ (1998) Effect of lung volume reduction surgery on neuromechanical coupling of the diaphragm. *Am J Respir Crit Care Med* 157: 475–483

48 Criner G, Cordova FG, Leyenson V, Roy B, Travaline J, Sudarshan S, O'Brien G, Kuzma AM, Furukawa S (1998) Effect of lung volume reduction surgery on diaphragm strength. *Am J Respir Crit Care Med* 157: 1578–1585

49 Wanke T, Merkle M, Formanek D, Zifko U, Wieselthaler G, Zwick H, Klepetko W, Burghuber OC (1994. Effect of lung transplantation on diaphragmatic function in patients with chronic obstructive pulmonary disease. *Thorax* 49: 459–464

50 Nocturnal Oxygen Therapy Trial Group (1980) Continuous or nocturnal oxygen therapy in hypoxaemic chronic obstructive lung disease. *Ann Intern Med* 93: 391–398

51 Fletcher EC, Donner CF, Midgren B, Zielinski J, Levi-Valensi P, Braghiroli A, Rida Z, Miller CC (1992) Survival in COPD patients with a daytime PaO2 greater than 60 mm Hg with and without nocturnal oxyhemoglobin desaturation. *Chest* 101: 649–655
52 Koulouris NG, Dimopoulou I, Valta P, Finkelstein R, Cosio MG, Milic-Emili J (1997) Detection of expiratory flow limitation during exercise in COPD patients. *J Appl Physiol* 82: 723–731

Computed tomography (CT) scans in COPD

Rachel C. Tennant[1], Trevor T. Hansel[1] and David M. Hansell[2]

[1]National Heart and Lung Institute (NHLI) Clinical Studies Unit, Royal Brompton Hospital, Fulham Road, London SW3 6HP, UK; [2]Department of Radiology, Royal Brompton Hospital, Fulham Road, London SW3 6HP, UK

Introduction

Structural information about the lungs has been obtainable for over one hundred years through imaging techniques. Cross-sectional computed tomography (CT) provides a great advance over conventional radiography, since it avoids the inevitable superimposition of structures that occurs with a projectional image, and CT allows resolution of structures as small as 200 μm. The digital nature of CT data also permits objective assessment of respiratory diseases including COPD. Sophisticated computer software has allowed the development of image processing techniques that may improve detection and quantification of COPD (Fig. 1). The development, over the past decade, of volumetric acquisition with spiral CT and 3D image reconstruction is an exciting development, the full value of which is only just being exploited. It is becoming apparent that CT can also be used to provide functional data, for example, looking at changes in airway calibre in response to drugs. An exciting development is that CT has been demonstrated to show very early smoking-related lung damage, and may be of use in monitoring resolution of such changes in response to treatment.

Plain chest radiography

Chest radiographs have not been found to be particularly useful in the diagnosis of smoking related lung disease. Abnormalities are readily detected when the disease is advanced, but changes in mild to moderate COPD tend to be subtle, and may be the vague "dirty CXR" of increased lung markings. These markings may consist of bronchial wall thickening and increased vascular markings, due to the presence of pulmonary hypertension. It has been suggested that this pattern correlates with chronic bronchitis, however many bronchitics have normal chest radiographs, and increased markings do not consistently represent disease [1].

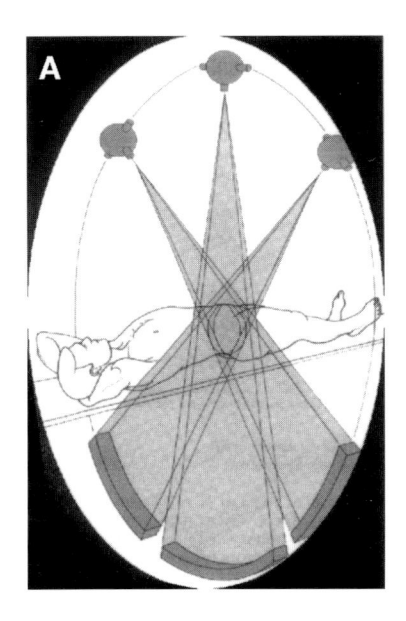

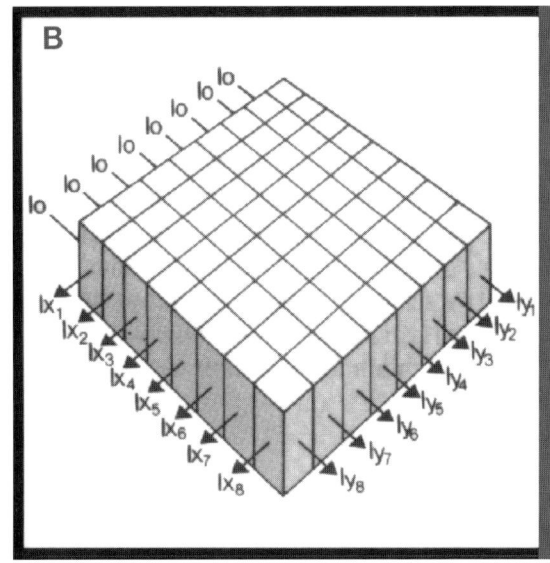

Figure 1
A. Principle of spiral (helical) CT. This modern form of CT facilitates three-dimensional imaging of airways, parenchyma and vessels through continuous scanning while the patient is moved through the CT gantry.
B. CT measures the attenuation of X-rays in Hounsfield Units (HU). The dimension of a voxel is in terms of length by width by thickness.

Changes due to emphysema manifest as hyperinflation and loss of vascular markings. Hyperinflation is demonstrated by flattened hemidiaphragms, a narrow cardiac silhouette, increased lung volumes and an increased retrosternal airspace on lateral view. Clinico-pathological correlations have provided varying results, with detection rates of emphysema on CXR between 65–80%, largely depending on the criteria used to make the radiological diagnosis. Sensitivity is improved when signs are used that indicate hyperinflation, as well as signs showing deficiency in the vascular pattern [2, 3].

Technical aspects

Multiple X-rays are passed through the patient in a fan-shaped beam, as the CT gantry makes a 360° rotation, and are detected on the opposite side of the patient (Fig. 1). Production of CT images depends on variable absorption (attenuation) of X-rays by different densities of tissue. An attenuation value is calculated for each defined volume of tissue (voxel), which corresponds to a pictorial element of the CT image (pixel). Attenuation values are expressed in Hounsfield units (HU). The value 0 corresponds to the attenuation of X-rays in water, whilst −1000 corresponds to the attenuation of X-rays in air (where they are absorbed least). Tissue more dense than water has an attenuation value of up to +3000 HU.

Conventional CT images display information as cross-sectional slices. The X-ray beam is collimated (focused) in 10 mm sections in the transverse plane. High resolution CT (HRCT) refers to scans acquired using thinner sections, usually 1 mm collimations taken at 10 or 20 mm intervals. This technique has the advantage of allowing visualisation of the lung parenchyma in much greater detail than is possible with conventional CT.

The development of CT technology over the past decade means that data from the entire volume of the thorax can be now be acquired in a single breath hold. The gantry of a "spiral" CT scanner rotates around the patient, and multiple detectors enable the rapid acquisition of data. Image processing techniques (multi-planar reformation) allow the presentation of images in any plane.

CT features of COPD

CT has been employed to identify the large range of airway, parenchymal and vascular pathology that may be present in COPD (see Table 1).

Emphysema

CT features of emphysema include decreased lung density ("black lung"), pulmonary vascular pruning and distortion and evidence of bulla formation [4]. Centrilobular, or smoking-related emphysema is characterised by areas of hypoattenuation, less than 1 cm in diameter detected at the centre of the secondary pulmonary lobule (Fig. 2). The areas of hypoattenuation do not have visible walls in contrast to cystic conditions. Another distinguishing feature is that centrilobular emphysematous lesions usually have an arteriole at the centre of the affected area. Centrilobular emphysema is most extensive in the upper lobes. In advanced cases, widespread damage is detected as larger areas of hypoattenuation without intervening normal lung tissue.

Panlobular emphysema, in α_1-antiprotease deficiency, predominantly affects the lower lobes. Uniform destruction of the pulmonary lobule is detectable as widespread areas of decreased lung density. The more rare paraseptal emphysema affects the distal part of the secondary pulmonary lobule, and is seen as areas of hypoattenuation adjacent to the pleura and septae.

Table 1 - Computed tomography (CT) features and corresponding lung pathology in COPD

CT feature	Pathophysiology
Bronchial wall thickening	Chronic bronchitis
Ground glass attenuation with nodular component	Respiratory bronchiolitis with interstitial lung disease (RBILD): Inflammation in and around bronchioles with accumulation of macrophages in alveoli.
Reticular pattern	Interstitial fibrosis
Mosaic attenuation pattern of lung parenchyma at end expiration	Obstruction of bronchioles causing regional underventilation with reduced perfusion
Multiple small round areas of low attenuation, with decreased Hounsfield units (HU) in the secondary pulmonary lobule. Several mm (< 1 cm) in diameter, surrounded by normal lung parenchyma. Upper lobe predominance.	Centrilobular emphysema: generally caused by cigarette smoking
Peripheral variant of centrilobular emphysema. Areas of low attenuation in the subpleural regions of the lung and adjacent to interlobular septa and pulmonary vessels.	Paraseptal emphysema: prone to lead to bulla and pneumothorax
Uniform destruction of the secondary pulmonary lobule gives homogeneous "black lung". Paucity of vascular markings. Diffuse distribution in lower lobes.	Panlobular emphysema: may be associated with α_1-antitrypsin deficiency
Thin walled, focal areas of avascularity. Bulla are > 1 cm in diameter, and tend to involve upper lobes.	Bulla
Vascular pruning, enlarged pulmonary artery Mosaic attenuation	Pulmonary hypertension

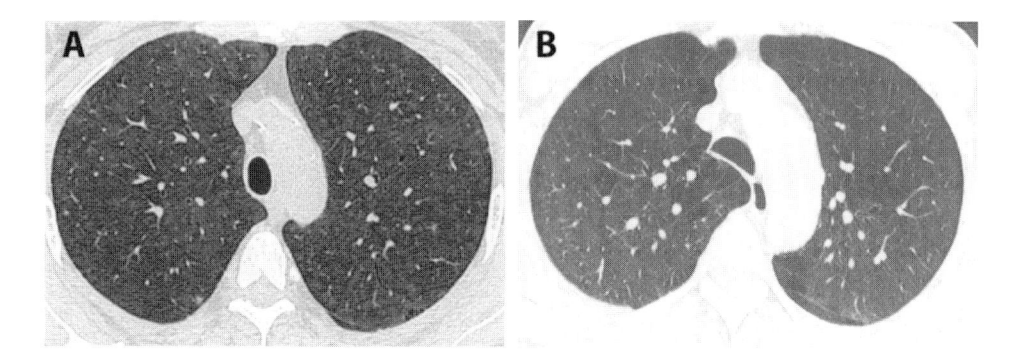

Figure 2
A. Respiratory bronchiolitis with interstitial lung disease (RBILD). Ill-defined white nodules are present, representing accumulation of macrophages in the interstitium and respiratory bronchiolitis.
B. Centrilobular emphysema. "Moth-eaten" appearance of lung with multiple patchy black areas (lower attenuation corresponding to lung destruction and centrilobular emphysema).

Quantification of emphysema

CT has been shown to be extremely useful in assessing the extent of emphysema. CT scans can be assessed in one of two ways. With visual assessment, which is subjective, an observer calculates the percentage of each scan involved by emphysema according to a predetermined scale, for example 0%, <25%, 25–50%, 50–75%, >75%. The digital nature of CT data also allows for an objective quantification of emphysema, based on attenuation values.

Several studies have compared the extent of emphysema seen on CT with the "gold standard" of histopathological assessment [5–13]. Some studies involved post-mortem specimens and others used resected lobes from patients undergoing surgery for lung cancer. Bergin and colleagues compared extent of emphysema assessed on conventional CT (10 mm collimation) by a visual scoring system with the pathological extent of emphysema [6]. The pathological extent of emphysema was graded by comparing the specimens to a panel of standards, and thus was not a truly quantitative method. A reasonable correlation was seen between the CT and pathology score ($r = 0.57$ for the resected lobe, $r = 0.63$ for the whole lung). It was found that CT was a better predictor of the pathological extent of emphysema than lung function tests. Other groups, using similar methods, found closer correlations when HRCT rather than conventional CT was performed [7–9]. However, when the pathological assessment was improved by using a more truly quantitative means of assessing the extent of emphysema it was found that visual assessment of CT consistently underestimated the extent of emphysema [9].

More recently the use of commercially-available software packages to aid image processing has made objective quantitative analysis of the degree of emphysema possible. Destruction of lung tissue in emphysema leads to affected areas of lung having increasingly negative attenuation values. This may be expressed in a variety of ways of which mean lung density and percentage of lung area with abnormally low attenuation values are most often used. There are several technical problems in the use of mean lung density as a measure of emphysema. Firstly it is important to define the boundaries of the lung parenchyma, and pixels towards the edge may contain adjoining tissue of higher density than lung, such as bone or muscle. Secondly, although the density of the lung parenchyma falls in emphysema, this is only a small change, which can be masked by variations in the pulmonary vasculature. Finally, lung density increases with expiration, as the air is expelled from the lungs, so alterations in readings may be found, if the patient is unable to breath hold at full inspiration, whilst images are obtained.

Early work demonstrated a shift in the frequency distribution curves of attenuation numbers towards more negative values in emphysema [10]. It was subsequently found that the lowest fifth percentile of the frequency histogram was significantly associated with a microscopic index quantifying emphysema [14]. This method can underestimate emphysema if it is associated with other disorders. It is therefore better to define a threshold attenuation value that distinguishes normal from emphysematous lung tissue.

Müller and colleagues proposed a method of image processing using a software package ("density mask") that quantifies emphysema by defining the percentage of lung occupied by areas of abnormally low attenuation [11]. The programme highlights voxels within a given density range (e.g., < −910 HU) and the percentage area of lung occupied by these voxels is automatically calculated. In a pilot study comparisons were made between conventional CT images (10 mm collimations, with contrast), and resected specimens, visually graded for severity, to calculate the optimal attenuation levels at which to set the density mask. Using these levels (< −920, −910 and −900 HU), the percentage area with abnormally low attenuation was assessed in 28 patients, and correlated with both the visual score for emphysema on the CT images, and the histological grading. Mean lung densities were also calculated. The results showed an excellent correlation between the density mask percentage area < −910 HU and the histological severity score ($r = 0.89$). There was a much weaker, although still significant, correlation between mean lung density and histology score ($r = 0.46$).

Quantitative analysis was taken an important step further by comparing CT data with pathology specimens assessed with a computer aided system that calculated the percentage area of emphysema in transverse macroscopic sections [12]. This increased the accuracy of the histological "gold standard" calculations. CT scans were quantified by calculating the percentage area occupied by attenuation numbers below thresholds ranging from −900 to −970 HU. A threshold of less than −950 HU gave a valid index of the extent of pulmonary emphysema. This study used high resolution rather than conventional CT, and contrast was not given, accounting for the difference found in the optimal threshold.

A further study compared CT images with a microscopic analysis of the lung resections [13]. Again, in this study the histology data was truly quantitative, with microscopic measurements reflecting the size of the alveoli and the alveolar ducts. There was a significant correlation between microscopic extent of emphysema and area with attenuation numbers below −950 HU ($r = 0.7$). Further correlation was made with pulmonary function tests. Correlations were seen with measures of airway obstruction (FEV_1, FEV_1/FVC) and air trapping (RV%, TLC%, RV/TLC%), but the closest correlations were with measure of gas transfer (DLCO, DLCO% predicted, DLCO/VA, DLCO%/VA).

Attempts have been made to correlate the extent of emphysema calculated by the "density mask" technique with lung function tests. Inverse correlations are seen with indices of airflow ($r = 0.51 - 0.72$) and of diffusing capacity ($r = 0.53$) [15]. While these correlations are not particularly tight, they are similar to those seen when lung function test results are correlated with pathological specimens, reflecting the fact that there are limitations in evaluating emphysema by lung function testing alone.

The studies described so far have involved CT scans taken at full inspiration (total lung capacity). It has been found that expiratory scans (taken at functional residual capacity) correlate more closely with diffusing capacity than do inspiratory images. It seems that permanent enlargement of the airspaces is better demonstrated on expiratory scans, and that inspiratory scans overdiagnose emphysema by detecting areas of hyperinflated lung that do not represent emphysema [16, 17].

The correlation between the degree of emphysema detected by CT and by pulmonary function testing is not particularly close. It has been shown that even though emphysema is usually more extensive in the upper lobes, it is the extent of lower lobe emphysema that correlates most closely with lung function results [18]. This reflects the fact that the contribution of the upper lobes to pulmonary function testing is small. Extensive emphysematous damage can occur before it is detected. On the other hand, CT may detect early emphysema that does not have functional significance.

Airways disease

While CT has been extensively studied in the assessment of emphysema, less attention has been paid to the imaging of airways disease. Emphysema is defined in histopathological terms, and so readily lends itself to detection by a technique that examines structure. Chronic bronchitis, however, is generally defined in clinical terms and so CT may not necessarily detect any morphological abnormalities.

Thickening of the walls of large airways can, however, be more readily detected in smokers using CT than by plain chest radiography. In one series, studying 175 healthy volunteers, bronchial wall thickening was detected in 33% of smokers on CT, but not in any cases by plain CXR [19]. In another study, calculation of airway wall thickness of the apical right upper lobe bronchus contributed to the prediction of lung function abnormalities when taken together with calculation of emphysematous involvement on CT [20].

There are considerable technical difficulties to be overcome when using CT to assess bronchial wall thickness and intra-luminal area. Several investigators have used phantom models of airways to assess errors in measurements using a variety of protocols [21]. In the main, it appears that accuracy is best in airways of larger diameter. Studies in animals [22], and in asthmatic volunteers [23, 24] have shown that CT has the potential to be of diagnostic use in the assessment of airways disease, but further investigations in COPD are necessary.

Early changes in smokers' lungs

The first part of the lung to be affected by smoking is the respiratory bronchiole. Histologically this is detected as mild chronic inflammation with accumulation of pigmented macrophages within the respiratory bronchioles and adjacent alveoli [25]. These histological changes are not usually associated with clinical symptoms, although in some cases more extensive involvement leads to a syndrome of cough, dyspnoea and a restrictive lung defect, known as respiratory bronchiolitis-interstitial lung disease (RB-ILD) (Fig. 2).

Damage to the small airways can be detected with lung function tests. Several investigators have attempted to identify sensitive tests of small airways function (MMEF, single breath N2 test, closing capacity) that may be used to monitor progression of smoking-related lung disease. Unfortunately it has been found that the reproducibility of the majority of these tests is low, and, in the main, abnormalities of these tests do not appear to predict the subsequent development of clinically significant airflow obstruction [26–28].

CT scanning can detect changes due to respiratory bronchiolitis; early signs of smoking-related damage include parenchymal nodules, areas of "ground glass" attenuation, and bronchial abnormalities. Significant differences in the frequency of these signs are found between current, ex- and non-smokers. They are more readily detected by HRCT than by conventional CT. In one study, HRCT detected parenchymal micronodules in 27% of smokers, but in only 4% of ex-smokers and no non-smokers ($p < 0.001$). The figures for areas of ground glass attenuation were 21%, 4% and 0% ($p = 0.001$), and for areas of emphysema 21%, 8% and 0% ($p < 0.001$) [19].

Pathological correlations demonstrate that parenchymal nodules represent brochiolectases with peribronchiolar fibrosis, and that ground glass attenuation indicates the presence of an inflammatory alveolar infiltrate [29].

A recent longitudinal study, carried out over a mean period of 5.5 years, has demonstrated the evolution of changes in smokers' lungs [30]. Persistent smokers were found to have an increase from baseline in the extent of emphysema and ground glass attenuation seen on HRCT. This was correlated with a significantly faster rate of decline in FEV_1 than in ex- and non-smokers. Subjects with micronodules seen at baseline were found to have no change, an increased number of micronodules or replacement of the nodules with areas of emphysema. This is the first study to demonstrate the evolution of emphysema from inflammatory lesions on CT to tissue destruction. It raises the possibility of identification of susceptible individuals at an early stage and the hope that intervention with smoking cessation may prevent progression of the disease [30].

Assessment prior to surgery

Surgical treatments for emphysema have been employed over several decades [31]. It is well established that resection of giant bullae leads to clinical improvement, and over the last twenty years single lung transplantation has become an available option to a few younger patients with terminal disease. More recently, the lung volume reduction procedure has come back in to vogue [31, 32–35]. Clearly, performing surgery on patients with severe lung disease is an alarming prospect, and careful patient selection is crucial. One of the major clinical applications of CT in COPD is in the pre-operative assessment of surgical candidates.

In the case of bullous emphysema, CT is of benefit in establishing the size of giant single bullae, and in identifying adjacent compressed lung. CT can be helpful in identifying the best site for the surgical approach, enabling mini-thoracotomy and reducing the morbidity associated with a larger wound [36].

Lung volume reduction surgery, originally proposed by Brantigan in the 1950s, was reintroduced by Cooper and colleagues in 1995 [37, 38]. The rationale behind this procedure is to re-establish normal lung mechanics by reducing hyperinflation, through resection of areas of diseased tissue. Cooper originally selected subjects on the basis of severe dyspnoea, increased total lung capacity, and a pattern of emphysema that included "target" regions of severe destruction, hyperinflation and poor perfusion.

It has since been found that the patients who derive most benefit have predominantly upper lobe disease that is heterogeneous, with the worst affected areas in the outer rind of the lung [39]. Many patients undergo ventilation-perfusion scintigraphy as well as CT imaging pre-operatively, however it has been shown that CT scanning is the most useful imaging modality and that scintigraphy has a limited role in the prediction of outcome [40].

Refinement of the selection criteria for this procedure is important since it has become apparent that while some patients derive great benefit, the operation is by no means a success for all. The National Emphysema Treatment Trial (NETT) in America is currently trying to address this issue [41]. Early results have shown that there is an unacceptable mortality risk for patients with a FEV_1 of less than 20% of predicted in association with either heterogeneous emphysema demonstrated on CT or a carbon monoxide diffusing capacity of less than 20% predicted [33]. The use of quantitative CT measurements to assess the extent and distribution of emphysematous lesions may prove useful in further refining the selection criteria for LVRS candidates in future.

Detection of early disease

CT is more sensitive than spirometry for the detection of early emphysema. There is a role for HRCT in establishing whether emphysema is the cause of dyspnoea in subjects with normal spirometry. Such patients may have an isolated gas transfer defect as the only lung function abnormality, however evidence of early emphysema can clearly be shown on CT [42].

Co-existent bronchiectasis and malignancies

CT is of use in smokers presenting with persistent productive cough to distinguish between COPD and bronchiectasis. In one investigation, 29% of subjects diagnosed with COPD in the community were found to have bronchiectasis on CT chest [43].

CT can also be used to screen for co-existent malignancy, a not infrequent occurrence in this patient group.

Monitoring progress

HRCT may be sufficiently sensitive to monitor longitudinal changes in the extent of emphysema. It has been shown that annual changes in lung density and percentage of low attenuation area are detectable in ex-smokers and current smokers with inspiratory HRCT. Annual increases in these parameters are greater than in a group of never smokers, however increases in airspace size due to normal ageing are detectable over five years [44]. Annual increases in extent of emphysema can also be detected using quantitative CT in patients with emphysema due to $α_1$-antitrypsin deficiency [45].

In another study, progression from pre-emphysematous lesions (micronodules) to emphysema was detectable in a group of smokers over five years. Only a small number of subjects gave up smoking over this period, but analysis of individual cases showed some resolution of pre-emphysematous lesions [30].

These findings raise the hope that CT may be a useful tool with which to monitor response to treatment in drug trials. It appears that subtle changes in disease extent may be more readily detected with CT imaging than with pulmonary function testing. Two on-going trials are using this approach. In a pilot study of α_1-antitrypsin augmentation therapy it was calculated that 130 patients would need to be recruited to demonstrate a significant effect using quantitative CT [46]. If annual decline in FEV_1 was used as the primary outcome measure then 550 subjects would be required. Change in extent of disease assessed by quantitative CT is also being used as an outcome measure in a study of all-*trans*-retinoic acid (ATRA) therapy [47].

Conclusion

The evidence suggests that CT is complementary to pulmonary function testing in the assessment of patients with COPD. The detection of early disease with CT may, however, be a useful motivational aid to encourage smoking cessation in apparently healthy smokers before severe damage is done. In addition, the ability to quantitate the extent of emphysema with CT means that it may be a powerful tool in monitoring response to therapies. The ability of CT to detect early signs of smoking-related damage, including respiratory bronchiolitis and alveolar infiltrates, may enable identification of early disease that may be reversible by novel anti-inflammatory therapies.

References

1   Guckel C and Hansell DM (1998) Imaging the 'dirty lung' – has high resolution computed tomography cleared the smoke? *Clin Radiol* 53: 717–722
2   Pratt PC (1987) Role of conventional chest radiography in diagnosis and exclusion of emphysema. *Am J Med* 82: 998–1006
3   Morgan MD (1992) Detection and quantification of pulmonary emphysema by computed tomography: a window of opportunity. *Thorax* 47: 1001–1004
4   Newell JD Jr (2002) CT of emphysema. *Radiol Clin North Am* 40: 31–42
5   Foster WL Jr, Pratt PC, Roggli VL, Godwin JD, Halvorsen RA, Putman CE (1986) Centrilobular emphysema: CT-pathologic correlation. *Radiology* 159: 27–32
6   Bergin C, Muller N, Nichols DM, Lillington G, Hogg JC, Mullen B, Grymaloski MR, Osborne S, Pare PD (1986) The diagnosis of emphysema. A computed tomographic-pathologic correlation. *Am Rev Respir Dis* 133: 541–546
7   Hruban RH, Meziane MA, Zerhouni EA, Khouri NF, Fishman EK, Wheeler PS, Dumler JS, Hutchins GM (1987) High resolution computed tomography of inflation-fixed

lungs. Pathologic-radiologic correlation of centrilobular emphysema. *Am Rev Respir Dis* 136: 935–940

8   Kuwano K, Matsuba K, Ikeda T, Murakami J, Araki A, Nishitani H, Ishida T, Yasumoto K, and Shigematsu N (1990) The diagnosis of mild emphysema. Correlation of computed tomography and pathology scores. *Am Rev Respir Dis* 141: 169–178

9   Miller RR, Muller NL, Vedal S, Morrison NJ, Staples CA (1989) Limitations of computed tomography in the assessment of emphysema. *Am Rev Respir Dis* 139: 980–983

10 Hayhurst MD, MacNee W, Flenley DC, Wright D, McLean A, Lamb A, Wightman AJ, Best J (1984) Diagnosis of pulmonary emphysema by computerised tomography. *Lancet* 2: 320–322

11 Muller NL, Staples CA, Miller RR, Abboud RT (1988) "Density mask". An objective method to quantitate emphysema using computed tomography. *Chest* 94: 782–787

12 Gevenois PA, de Maertelar V, de Vuyst P, Zanen J, and Yernault JC (1995) Comparison of computed density and macroscopic morphometry in pulmonary emphysema. *Am J Respir Crit Care Med* 152: 653–657

13 Gevenois PA, de Vuyst P, de Maertelar V, Zanen J, Jacobovitz D, Cosio MG, Yernault JC (1996) Comparison of computed density and microscopic morphometry in pulmonary emphysema. *Am J Respir Crit Care Med* 154: 187–192

14 Gould GA, MacNee W, McLean A, Warren PM, Redpath A, Best JJ, Lamb D, and Flenley DC (1988) CT measurements of lung density in life can quantitate distal airspace enlargement – an essential defining feature of human emphysema. *Am Rev Respir Dis* 137: 380–392

15 Kinsella M, Muller NL, Abboud RT, Morrison NJ, DyBuncio A (1990) Quantitation of emphysema by computed tomography using a "density mask" program and correlation with pulmonary function tests. *Chest* 97: 315–321

16 Knudson RJ, Standen JR, Kaltenborn WT, Knudson DE, Rehm K, Habib MP, Newell JD (1991) Expiratory computed tomography for assessment of suspected pulmonary emphysema. *Chest* 99: 1357–1366

17 Heremans A, Verschakelen JA, Van Fraeyenhoven L, Demedts M (1992) Measurement of lung density by means of quantitative CT scanning. A study of correlations with pulmonary function tests. *Chest* 102: 805–811

18 Gurney JW, Jones KK, Robbins RA, Gossman GL, Nelson KJ, Daughton D, Spurzem JR, Rennard SI (1992) Regional distribution of emphysema: correlation of high-resolution CT with pulmonary function tests in unselected smokers. *Radiology* 183: 457–463

19 Remy-Jardin M, Remy J, Boulenguez C, Sobaszek A, Edme JL, Furon D (1993) Morphologic effects of cigarette smoking on airways and pulmonary parenchyma in healthy adult volunteers: CT evaluation and correlation with pulmonary function tests. *Radiology* 186: 107–115

20 Nakano Y, Muro S, Sakai H, Hirai T, Chin K, Tsukino M, Nishimura K, Itoh H, Pare PD, Hogg JC, Mishima M (2000) Computed tomographic measurements of airway dimensions and emphysema in smokers. Correlation with lung function. *Am J Respir Crit Care Med* 162: 1102–1108

21 Webb WR, Gamsu G, Wall SD, Cann CE, Proctor E (2002) CT of a bronchial phantom: factors affecting appearance and size measurements. *Invest Radiol* 19, 394–398
22 Brown RH, Herold CJ, Hirshman CA, Zerhouni EA, Mitzner W (1991) *In vivo* measurements of airway reactivity using high-resolution computed tomography. *Am Rev Respir Dis* 144: 208–212
23 McNitt-Gray MF, Goldin JG, Johnson TD, Tashkin DP, Aberle DR (1997) Development and testing of image-processing methods for the quantitative assessment of airway hyperresponsiveness from high-resolution CT images. *J Comput Assist Tomogr* 21: 939–947
24 Mclean AN, Sproule MW, Cowan MD Thomson NC (1998) High resolution computed tomography in asthma. *Thorax* 53: 308–314
25 Niewoehner DE, Kleinerman J, Rice DB (1974) Pathologic changes in the peripheral airways of young cigarette smokers. *N Engl J Med* 291: 755–758
26 Tattersall SF, Benson MK, Hunter D, Mansell A, Pride NB, Fletcher CM (1978) The use of tests of peripheral lung function for predicting future disability from airflow obstruction in middle-aged smokers. *Am Rev Respir Dis* 118: 1035–1050
27 Stanescu DC, Rodenstein DO, Hoeven C, Robert A (1987) "Sensitive tests" are poor predictors of the decline in forced expiratory volume in one second in middle-aged smokers. *Am Rev Respir Dis* 135: 585–590
28 Stanescu D, Sanna A, Veriter C, Robert A (1998) Identification of smokers susceptible to development of chronic airflow limitation: a 13-year follow-up. *Chest* 114: 416–425
29 Remy-Jardin M, Remy J, Gosselin B, Becette V, Edme JL (1993) Lung parenchymal changes secondary to cigarette smoking: pathologic-CT correlations. *Radiology* 186: 643–651
30 Remy-Jardin M, Edme JL, Boulenguez C, Remy J, Mastora I, Sobaszek A (2002) Longitudinal follow-up study of smoker's lung with thin-section CT in correlation with pulmonary function tests. *Radiology* 222: 261–270
31 Cooper, JD (1997) The history of surgical procedures for emphysema. *Ann Thorac Surg* 63: 312–319
32 Geddes D, Davies M, Koyama H, Hansell DM, Pastorino U, Pepper J, Agent P, Cullinan P, MacNeill SJ, Goldstraw P (2000) Effect of lung-volume-reduction surgery in patients with severe emphysema. *N Engl J Med* 343: 239–245
33 National Emphysema Treatment Trial Research Group (2001) Patients at high risk of death after lung volume reduction surgery. *N Engl J Med* 345: 1075–1083
34 Fishman A, Martinez F, Naunheim K, Piantadosi S, Wise R, Ries A, Weinmann G, Wood DE (2003) National Emphysema Treatment Trial Research Group. A randomized trial comparing lung-volume-reduction surgery with medical therapy for severe emphysema. *N Engl J Med* 348: 2059–2073
35 Ramsey SD, Berry K, Etzioni R, Kaplan RM, Sullivan SD, Wood DE (2003) National Emphysema Treatment Trial Research Group. Cost effectiveness of lung-volume-reduction surgery for patients with severe emphysema. *N Engl J Med* 348: 2092–2102

36 Morgan MD, Denison DM, Strickland B (2002) Value of computed tomography for selecting patients with bullous lung disease for surgery. *Thorax* 41: 855
37 Cooper JD, Trulock EP, Triantafillou AN, Patterson GA, Pohl MA, Deloney PA, Sundaresen RS, Roper CL (2002) Bilateral pneumonectomy (volume reduction) for chronic obstructive pulmonary disease. *J Thorac Cardiovasc Surg* 109: 106–116
38 Cooper JD, Patterson GA, Sundaresan RS, Trulock EP, Yusen RD, Pohl MS, Lefrak SS (1996) Results of 150 consecutive bilateral lung volume reduction procedures in patients with severe emphysema. *J Thorac Cardiovasc Surg* 112: 1319–1329
39 Nakano Y, Coxson HO, Bosan S, Rogers RM, Sciurba FC, Keenan RJ, Walley KR, Pare PD, Hogg JC (2001) Core to rind distribution of severe emphysema predicts outcome of lung volume reduction surgery. *Am J Respir Crit Care Med* 164: 2195–2199
40 Thumheer R, Engel H, Weder W, Stammberger U, Laube I, Russi EW, Bloch K (2002) Role of lung perfusion scintigraphy in relation to chest computed tomography and pulmonary function in the evaluation of candidates for lung volume reduction surgery. *Am J Respir Crit Care Med* 159: 301–310
41 The National Emphysema Treatment Trial Research Group (1999) Rationale and design of The National Emphysema Treatment Trial: a prospective randomized trial of lung volume reduction surgery. *Chest* 116: 1750–1761
42 Klein JS, Gamsu G, Webb WR, Golden JA, Muller NL (1992) High-resolution CT diagnosis of emphysema in symptomatic patients with normal chest radiographs and isolated low diffusing capacity. *Radiology* 182: 817–821
43 O'Brien C, Guest PJ, Hill SL, Stockley RA (2000) Physiological and radiological characterisation of patients diagnosed with chronic obstructive pulmonary disease in primary care. *Thorax* 55: 635–642
44 Soejima K, Yamaguchi K, Kohda E, Takeshita K, Ito Y, Mastubara H, Oguma T, Inoue T, Okubo Y, Amakawa K et al (2000) Longitudinal follow-up study of smoking-induced lung density changes by high-resolution computed tomography. *Am J Respir Crit Care Med* 161: 1264–1273
45 Dowson LJ, Guest PJ, Stockley RA (2001) Longitudinal changes in physiological, radiological, and health status measurements in α_1-antitrypsin deficiency and factors associated with decline. *Am J Respir Crit Care Med* 164: 1805–1809
46 Dirksen A, Dijkman JH, Madsen F, Stoel B, Hutchison DCS, Ulrik CS, Skovgaard LT, Kok-Jensen A, Rudolphus A, Seersholm N et al (1999) A randomized clinical trial of α_1-antitrypsin augmentation therapy. *Am J Respir Crit Care Med* 160: 1468–1472
47 Mao JT, Goldin JG, Dermand J, Ibrahim G, Brown MS, Emerick A, McNitt-Gray MF, Gjertson DW, Estrada F, Tashkin DP et al (2002) A pilot study of all-trans-retinoic acid for the treatment of human emphysema. *Am J Respir Crit Care Med* 165: 718–723

Oxidative stress in COPD

Peter J. Barnes

National Heart and Lung Institute, Imperial College, Dovehouse Street, London SW3 6LY, UK

Introduction

Oxidative stress occurs when reactive oxygen species (ROS) are produced in excess of the antioxidant defence mechanisms and result in harmful effects, including damage to lipids, proteins and DNA. There is increasing evidence that oxidative stress is an important feature in inflammatory airway diseases, including chronic obstructive pulmonary disease (COPD) [1–3]. This area of research has received new impetus by the recent development of several techniques for measuring oxidative stress in the lungs [4]. However, the pathophysiological role of oxidative stress in airway diseases will only be firmly established when more potent antioxidants become available for clinical use.

Formation

Inflammatory and structural cells that are activated in the airways of patients with COPD produce ROS, including, neutrophils, eosinophils, macrophages, and epithelial cells [2]. Superoxide anions ($O_2^{\cdot-}$) are generated by NADPH oxidase and this is converted to hydrogen peroxide (H_2O_2) by superoxide dismutases. H_2O_2 is then dismuted to water by catalase. $O_2^{\cdot-}$ and H_2O_2 may interact in the presence of free iron to form the highly reactive hydroxyl radical ($\cdot OH$). $O_2^{\cdot-}$ may also combine with NO to form peroxynitrite, which also generates $\cdot OH$ [5]. Oxidative stress leads to the oxidation of arachidonic acid and the formation of a new series of prostanoid mediators called isoprostanes, which may exert significant functional effects [6], including bronchoconstriction and plasma exudation [7–9].

Granulocyte peroxidases, such as myeloperoxidase in neutrophils, play an important role on oxidative stress. In neutrophils, H_2O_2 generated from O_2^- is metabolised by myeloperoxidase in the presence of chloride ions to hypochlorous acid which is a strong oxidant. Myeloperoxidase is also able to nitrate tyrosine residues, as can peroxynitrite [10–12].

Antioxidants

Oxidative stress describes an imbalance between ROS and antioxidants. The normal production of oxidants is counteracted by several antioxidant mechanisms in the human respiratory tract [13]. The major intracellular antioxidants in the airways are catalase, SOD and glutathione, formed by the enzyme γ-glutamyl cysteine synthetase, and glutathione synthetase. Thioredoxin may also be an important regulator of oxidative stress, but little is known about its role in the airways [14]. Oxidant stress activates the inducible enzyme heme oxygenase-1 (HO-1), converting heme and hemin to biliverdin with the formation of carbon monoxide (CO) [15]. Biliverdin is converted *via* bilirubin reductase to bilirubin, which is a potential antioxidant. HO-1 is widely expressed in human airways [16] and CO production is increased in COPD (see below). HO-1 activation may be a marker of oxidative stress but may also play a protective role in the airways.

In the lung intracellular antioxidants are expressed at relatively low levels and are not induced by oxidative stress, whereas the major antioxidants are extracellular [17]. Extracellular antioxidants, particularly glutathione peroxidase, are markedly up-regulated in response to cigarette smoke and oxidative stress. The glutathione system is the major antioxidant mechanism in the airways. There is a high concentration of reduced glutathione (GSH) in lung epithelial lining fluid [13] and concentrations are increased higher in cigarette smokers. Extracellular glutathione peroxidase (eGPx) is an important antioxidant in the lungs and may be secreted by epithelial cells and macrophages, particularly in response to cigarette smoke or oxidative stress [18]. eGPx inactivates H_2O_2 and O_2^-, but may also reactivate nitrogen species [17]. Extracellular antioxidants also include the dietary antioxidants vitamin C (ascorbic acid) and vitamin E (α-tocopherol), uric acid, lactoferrin and extracellular superoxide dismutase (SOD3). SOD3 is highly expressed in human lung, but its role in COPD is not yet clear [19].

Effects on airways

ROS have several effects on the airways, which would have the effect of increasing the inflammatory response (Fig. 1). These effects may be mediated by direct actions of ROS on target cells in the airways, but may also be mediated indirectly *via* activation of signal transduction pathways and transcription factors and *via* the formation of oxidised mediators such as isoprostanes and hydroxyl-nonenal.

Effects on signal transduction pathways and transcription factors

ROS activate the transcription factor nuclear factor-κB (NF-κB), which switches on

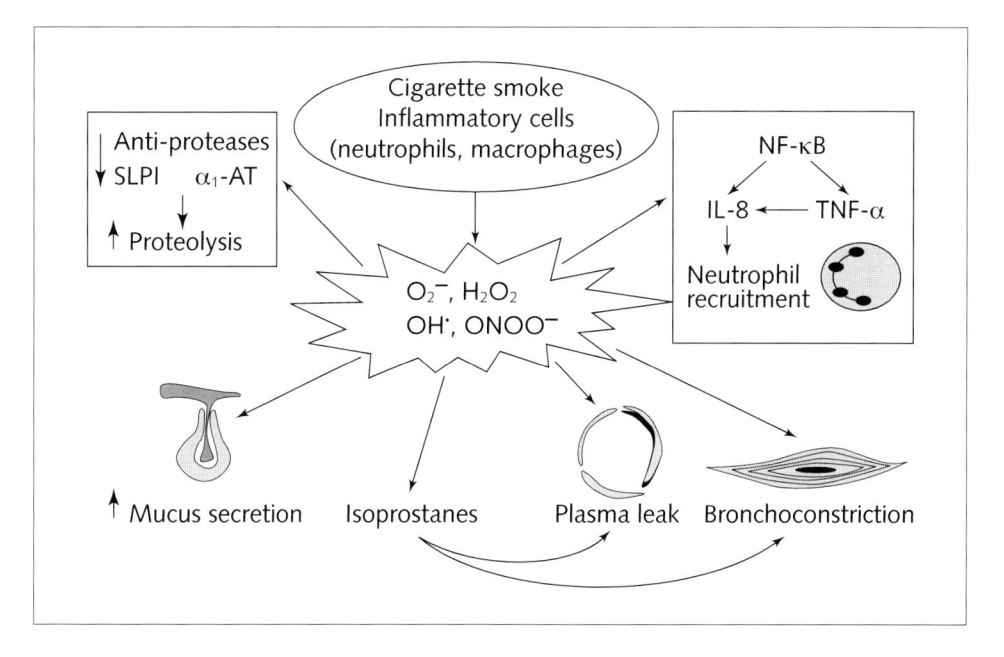

Figure 1
Oxidative stress in COPD.

multiple inflammatory genes [20, 21]. The molecular pathways by which oxidative stress activates NF-κB have not been fully elucidated, but there are several redox-sensitive steps in the activation pathway [22]. Another transcription factor that activates inflammatory genes is activator protein-1 (AP-1) which is a heterodimer of Fos and Jun proteins. As with NF-κB there are several redox-sensitive steps in the activation pathway [23]. Exogenous oxidants may also be important in worsening airway disease. Cigarette smoke, ozone and, to a lesser extent, nitrogen dioxide, impose an oxidative stress on the airways [24].

Oxidants also activate mitogen-activated protein (MAP) kinase pathways. H_2O_2 is a potent activator of extracellular regulated kinases (ERK) and p38 MAP kinase pathways that regulate the expression of many inflammatory genes and survival in certain cells, and spreading of macrophages [25]. Indeed many aspects of macrophage function are regulated by oxidants through the activation of multiple kinase pathways [26].

Airway smooth muscle

H_2O_2 directly constricts airway smooth muscle *in vitro* and this effect is mediated partly *via* the release of prostanoids [27]. ROS may damage airway epithelium, resulting in increased epithelial shedding and increased bronchoconstrictor responses [28]. Superoxide dismutase results in a decreased reactivity of guinea-pig trachea *in vitro*, suggesting that baseline production of $O_2^{\cdot-}$ may increase reactivity [29]. *In vitro* H_2O_2 induces an increase in responsiveness of human airways [30]. Formation of peroxynitrite also increases airway responsiveness in guinea pigs *in vitro* and *in vivo* [31, 32], but its effect in human airways is not yet known. 8-isoprostane (or 8-epi-prostaglandin $F_{2\alpha}$), the predominant isoprostane formed by the non-enzymatic oxidation of arachidonic acid in humans, is a potent constrictor of animal and human airways *in vitro*, an effect that is largely mediated *via* thromboxane (TP) receptors [7].

Vessels

Little is known about the effects of ROS on the bronchial vasculature. $\cdot$OH potently induce plasma exudation in rodent airways [33]. 8-isoprostane is a potent inducer of plasma exudation in airways [8].

Mucus secretion

In rats, oxidative stress increases airway mucus secretion, an effect that is blocked by cyclo-oxygenase inhibitors [34]. The effect of oxidative stress may be mediated *via* the activation of epidermal growth factor receptors on submucosal glands [35]. Neutrophil elastase is a potent stimulant of mucus secretion and increases the expression of mucin genes (MUC5AC); its effects are inhibited by dimethylthiourea, a purported scavenger of $\cdot$OH [36].

Nerves

Allergen impairs the function of bronchodilator nerves in guinea pig airways *in vivo* by an effect that is blocked by superoxide dismutase (SOD), suggesting that $O_2^{\cdot-}$ may scavenge NO released from motor nerves [37]. In rat airways, oxidant stress increases cholinergic nerve-induced bronchoconstriction, an effect that may be due to oxidant damage of acetylcholinesterase [38]. 8-isoprostane also has direct effects of airway nerves [39].

Inflammatory effects

Oxidants activate NF-κB, which orchestrates the expression of multiple inflammatory genes that have increased expression in COPD, thereby amplifying the inflammatory response [20]. Many of the stimuli that activate NF-κB appear to do so *via* the formation of ROS, particularly H_2O_2. ROS activate NF-κB in an epithelial cell line [40] and increase the release of pro-inflammatory cytokines from cultured human airway epithelial cells [41]. Oxidative stress results in activation of histone acetyltransferase activity which opens up the chromatin structure and is associated with increased transcription of multiple inflammatory genes [42, 43].

Increased oxidative stress

There is considerable evidence for increased oxidative stress in COPD [1, 2]. Cigarette smoke itself contains a high concentration of ROS [44]. Inflammatory cells, such as activated macrophages and neutrophils, also generate ROS, as discussed above.

Epidemiological evidence indicates that reduced dietary intake of antioxidants may be a determinant of COPD and population surveys have linked a low dietary intake of the antioxidant ascorbic acid (vitamin C) with worse lung function [45, 46].

Exhaled markers of oxidative stress

There are several markers of oxidative stress that may be detected in the breath and several studies have demonstrated increased production of oxidants in exhaled air or breath condensates [4, 47, 48]. There is an increased concentration of H_2O_2 in exhaled breath condensate of patients with COPD, particularly during exacerbations [49, 50].

There is also an increase in the concentration of 8-iso prostaglandin $F_{2\alpha}$ (8-isoprostane) in exhaled breath condensate, which is found even in patients who are ex-smokers [51] and is increased further during acute exacerbations [52]. Isoprostane is also increased in the breath of normal smokers, but to a lesser extent than in COPD, suggesting that there is an exaggeration of oxidative stress in COPD. 8-isoprostane is similarly increased in the urine of patients with COPD and further increased during exacerbations [53].

The concentrations of exhaled carbon monoxide (CO) are also increased in patients with COPD [54]. However in smokers this is difficult to interpret because cigarettes contain CO, but this increase is also found in confirmed ex-smokers and is presumed to reflect increased activation of heme oxygenase-1 in response to oxidative stress. Exhaled ethane, a volatile marker of lipid peroxidation, is also increased in COPD and is correlated with the levels of exhaled CO [55]. Another marker of lipid peroxidation thiobarbituric acid reactive substances (TBARS) is also increased in exhaled breath condensate of patients with COPD [50].

The concentrations of nitric oxide (NO) in exhaled air are not increased in COPD patients to the same extent as in asthma [54, 56, 57], but increase during exacerbations [56, 58]. In normal smokers exhaled NO levels are lower than normal and this is related to the number of cigarettes smoked [59]. This may reflect consumption of NO gas due to its avid interaction with superoxide anions to form peroxynitrite, which is then converted to nitrate. In patients with COPD there is increased expression of inducible NO synthase in alveolar macrophages and the NO generated may form peroxynitrite which then nitrates tyrosine residues on proteins to form 3-nitrotyrosine adducts which may be detected by immunocytochemistry. There is a marked increase in 3-nitrotyrosine immunoreactivity in sputum macrophages in patients with COPD [60]. There is also an increase in nitrite and nitrate concentrations in exhaled breath condensates, which may also reflect peroxynitrite formation [61].

Other evidence for oxidative stress

There is also evidence for increased systemic markers of oxidative stress in patients with COPD as measured by biochemical markers of lipid peroxidation [62]. A specific marker lipid peroxidation 4-hydoxy-2-nonenal which forms adducts with basic amino acid residues in proteins can be detected by immunocytochemistry and has been detected in lungs of patients with COPD [63]. This signature of oxidative stress is localised to airway and alveolar epithelial cells, endothelial cells and neutrophils.

Role of oxidative stress in COPD

The increased oxidative stress in the airways of COPD patient may play an important pathophysiological role in the disease.

Amplification of inflammation

Oxidative stress may be a means of amplifying the inflammatory response in COPD. This may reflect the activation of transcription factors such as NF-κB and AP-1, which then induce a neutrophilic inflammation *via* increased expression of IL-8 and other CXC chemokines, TNF-α and MMP-9. NF-κB is activated in airways and alveolar macrophages of patients with COPD and is further activated during exacerbations [64, 65]. It is likely that oxidative stress is an important activation of this transcription factor in COPD patients.

Reduction in antiproteases

Oxidative stress may also impair the function of antiproteases such as α_1-antitrypsin and secretory leukoprotease inhibitor, and thereby accelerate the breakdown of elastin in lung parenchyma [66].

Corticosteroid resistance

Corticosteroids are much less effective in COPD than in asthma and do not reduce the progression of the disease [67–70]. In contrast to patients with asthma, those with COPD do not show any significant anti-inflammatory response to corticosteroids [71, 72]. Alveolar macrophages from patients with COPD show a marked reduction in responsiveness to the anti-inflammatory effects of corticosteroids, compared to cells from normal smokers and non-smokers [73].

Recent studies suggest that there may be a link between oxidative stress and the poor response to corticosteroids in COPD. Oxidative stress impairs binding of glucocorticoid receptors to DNA and the translocation of these receptors from the cytoplasm to the nucleus [74, 75]. Corticosteroids switch off inflammatory genes by recruiting histone deacetylase-2 (HDAC2) to the active transcription site and by deacetylating the hyperacetylated histones of the actively transcribing inflammatory gene, they are able to switch off its transcription and thus suppress inflammation [76, 77]. In cigarette smokers and patients with COPD there is a marked reduction in activity of HDAC and reduced expression of HDAC2 in alveolar macrophages [78] and an even greater reduction in HDAC2 expression in peripheral lung tissue [79]. This reduction in HDAC activity is correlated with reduced expression of inflammatory cytokines and a reduced response to corticosteroids. This may result directly or indirectly from oxidative stress and is mimicked by the effects of H_2O_2 in cell lines [79].

Apoptosis

Oxidative stress may also induce apoptosis in endothelial and epithelial cells. Apoptosis of type 1 pneumocytes may be contributory to the development of emphysema and this might be induced by cytotoxic T lymphocytes or by inhibition of vascular-endothelial growth factor receptors [80, 81]. ROS may induce apoptosis by activating the NF-κB pathway, by direct DNA damage *via* activation of poly-ADP-ribose and *via* the generation of 4-hydroxy-nonenal. Apoptosis signal-regulating kinase-1 is held in an inactive conformation by thioredoxin and when oxidised by ROS this triggers apoptotic pathways [82].

Antioxidants as therapy for COPD

In view of the persuasive evidence presented above that oxidative stress is important in the pathophysiology of COPD antioxidants are a logical approach to therapy [83, 84].

Several antioxidants have also been administered to patients with COPD to explore their effects on lung function. NAC was developed as a mucolytic agent but also acts as an antioxidant by increasing the formation of glutathione. Although small scale trials failed to demonstrate any clear clinical benefit, more recent meta-analyses have shown a small but significant clinical benefit in COPD, particularly in reducing exacerbations [85, 86]. This benefit is not shared by other mucolytics and is therefore likely to be due to the antioxidant effect of NAC. These results should encourage the development of more effective antioxidants in the future.

Currently available antioxidants are rather weak, but more potent drugs, including spin-trap antioxidants (nitrones) and stable glutathione analogues are currently in clinical development [87]. Inhibitors of inducible NO synthase may inhibit formation of peroxynitrite and may be of value in therapy in view of the potential detrimental role of peroxynitrite. Selective, potent and long-lasting inhibitors of iNOS are now in clinical development [88]. The selenium-containing antioxidant ebselen is reported to be effective as an efficient scavenger of peroxynitrite [89], but does not appear to be in clinical development.

References

1 Repine JE, Bast A, Lankhorst I (1997) Oxidative stress in chronic obstructive pulmonary disease. *Am J Respir Crit Care Med* 156: 341–357
2 Macnee W (2001) Oxidative stress and lung inflammation in airways disease. *Eur J Pharmacol* 429: 195–207
3 Henricks PA, Nijkamp FP (2001) Reactive oxygen species as mediators in asthma. *Pulm Pharmacol Ther* 14: 409–420
4 Kharitonov SA, Barnes PJ (2001) Exhaled markers of pulmonary disease. *Am J Respir Crit Care Med* 163: 1693–1772
5 Beckman JS, Koppenol WH (1996) Nitric oxide, superoxide, and peroxynitrite: the good, the bad, and the ugly. *Am J Physiol* 271: C1432–C1437
6 Morrow JD (2000) The isoprostanes: their quantification as an index of oxidant stress status *in vivo*. *Drug Metab Rev* 32: 377–385
7 Kawikova I, Barnes PJ, Takahashi T, Tadjkarimi S, Yacoub MH, Belvisi MG (1996) 8-epi-prostaglandin $F_{2\alpha}$, a novel non-cyclooxygenase derived prostaglandin, is a potent constrictor of guinea-pig and human airways. *Am J Respir Crit Care Med* 153: 590–596
8 Okazawa A, Kawikova I, Cui ZH, Skoogh BE, Lotvall J (1997) 8-Epi-PGF2alpha

induces airflow obstruction and airway plasma exudation *in vivo*. *Am J Respir Crit Care Med* 155: 436–441

9   Janssen LJ (2001) Isoprostanes: an overview and putative roles in pulmonary pathophysiology. *Am J Physiol Lung Cell Mol Physiol* 280: L1067–L1082

10 van der Vliet A, Eiserich JP, Shigenaga MK, Cross CE (1999) Reactive nitrogen species and tyrosine nitration in the respiratory tract: epiphenomena or a pathobiologic mechanism of disease? *Am J Respir Crit Care Med* 160: 1–9

11 Eiserich JP, Hristova M, Cross CE, Jones AD, Freeman BA, Halliwell B, van der Vliet A (1998) Formation of nitric oxide-derived inflammatory oxidants by myeloperoxidase in neutrophils. *Nature* 391: 393–397

12 Gaut JP, Byun J, Tran HD, Lauber WM, Carroll JA, Hotchkiss RS, Belaaouaj A, Heinecke JW (2002) Myeloperoxidase produces nitrating oxidants *in vivo*. *J Clin Invest* 109: 1311–1319

13 Cantin AM, Fells GA, Hubbard RC, Crystal RG (1990) Antioxidant macromolecules in the epithelial lining fluid of the normal human lower respiratory tract. *J Clin Invest* 86: 962–971

14 Nordberg J, Arner ES (2001) Reactive oxygen species, antioxidants, and the mammalian thioredoxin system. *Free Radic Biol Med* 31: 1287–1312

15 Choi AM, Alam J (1996) Heme oxygenase-1: function, regulation, and implication of a novel stress-inducible protein in oxidant-induced lung injury. *Am J Respir Cell Mol Biol* 15: 9–19

16 Lim S, Groneberg D, Fischer A, Oates T, Caramori G, Mattos W, Adcock J, Barnes PJ, Chung KF (2000) Expression of heme oxygenase isoenzymes 1 and 2 in normal and asthmatic airways. Effect of inhaled corticosteroids. *Am J Respir Crit Care Med* 162: 1912–1918

17 Comhair SA, Erzurum SC (2002) Antioxidant responses to oxidant-mediated lung diseases. *Am J Physiol Lung Cell Mol Physiol* 283: L246–L255

18 Avissar N, Finkelstein JN, Horowitz S, Willey JC, Coy E, Frampton MW, Watkins RH, Khullar P, Xu YL, Cohen HJ (1996) Extracellular glutathione peroxidase in human lung epithelial lining fluid and in lung cells. *Am J Physiol* 270: L173–L182

19 Bowler RP, Crapo JD (2002) Oxidative stress in airways: is there a role for extracellular superoxide dismutase? *Am J Respir Crit Care Med* 166: S38–S43

20 Barnes PJ, Karin M (1997) Nuclear factor-κB: a pivotal transcription factor in chronic inflammatory diseases. *New Engl J Med* 336: 1066–1071

21 Barnes PJ, Adcock IM (1998) Transcription factors and asthma. *Eur Respir J* 12: 221–234

22 Janssen-Heininger YM, Poynter ME, Baeuerle PA (2000) Recent advances towards understanding redox mechanisms in the activation of nuclear factor κB. *Free Radic Biol Med* 28: 1317–1327

23 Xanthoudakis S, Curran T (1996) Redox regulation of AP-1: a link between transcription factor signaling and DNA repair. *Adv Exp Med Biol* 387: 69–75

24 Devalia JL, Bayram H, Rusznak C, Calderon M, Sapsford RJ, Abdelaziz MA, Wang J,

Davies RJ (1997) Mechanisms of pollution-induced airway disease: *in vitro* studies in the upper and lower airways. *Allergy* 52: 45–51; discussion 57–8

25 Ogura M, Kitamura M (1998) Oxidant stress incites spreading of macrophages *via* extracellular signal-regulated kinases and p38 mitogen-activated protein kinase. *J Immunol* 161: 3569–3574

26 Forman HJ, Torres M (2002) Reactive oxygen species and cell signaling: respiratory burst in macrophage signaling. *Am J Respir Crit Care Med* 166: S4–S8

27 Rhoden KJ, Barnes PJ (1989) Effect of oxygen-derived free radicals on responses of guinea-pig tracheal smooth muscle *in vitro*. *Br J Pharmacol* 98: 325–330

28 Yukawa T, Read RC, Kroegel C, Rutman A, Chung KF, Wilson R, Cole PJ, Barnes PJ (1990) The effects of activated eosinophils and neutrophils on guinea pig airway epithelium *in vitro*. *Am J Respir Cell Mol Biol* 2: 341–354

29 de Boer J, Pouw FM, Zaagsma J, Meurs H (1998) Effects of endogenous superoxide anion and nitric oxide on cholinergic constriction of normal and hyperreactive guinea pig airways. *Am J Respir Crit Care Med* 158: 1784–1789

30 Hulsmann AR, Raatgeep HR, den Hollander JC, Stijnen T, Saxena PR, Kerrebijn KF, de Jongste JC (1994) Oxidative epithelial damage produces hyperresponsiveness of human peripheral airways. *Am J Respir Crit Care Med* 149: 519–525

31 Sadeghi-Hashjin G, Folkerts G, Henricks PAJ, Verheyen AK, van der Linde HJ, van Ark I, Coene A, Nijkamp FP (1996) Peroxynitirite induces airway hyperresponsiveness in guinea pigs *in vitro* and *in vivo*. *Am J Respir Crit Care Med* 153: 1697–1701

32 de Boer J, Meurs H, Flendrig L, Koopal M, Zaagsma J (2001) Role of nitric oxide and superoxide in allergen-induced airway hyperreactivity after the late asthmatic reaction in guinea-pigs. *Br J Pharmacol* 133: 1235–1242

33 Lei Y-H, Barnes PJ, Rogers DF (1996) Involvement of hydroxyl radicals in neurogenic airway plasma exudation and bronchoconstriction in guinea pigs *in vivo*. *Br J Pharmacol* 117: 449–454

34 Adler KB, Holden Stauffer WJ, Repine JE (1990) Oxygen metabolites stimulate release of high-molecular-weight glycoconjugates by cell and organ cultures of rodent respiratory epithelium *via* an arachidonic acid-dependent mechanism. *J Clin Invest* 85: 75–85

35 Takeyama K, Dabbagh K, Jeong SJ, Dao-Pick T, Ueki IF, Nadel JA (2000) Oxidative stress causes mucin synthesis *via* transactivation of epidermal growth factor receptor: role of neutrophils. *J Immunol* 164: 1546–1552

36 Fischer B, Voynow J (2000) Neutrophil elastase induces MUC5AC messenger RNA expression by an oxidant-dependent mechanism. *Chest* 117: 317S–320S

37 Miura M, Yamauchi H, Ichinose M, Ohuchi Y, Kageyama N, Tomaki M, Endoh N, Shirato K (1997) Impairment of neural nitric oxide-mediated relaxation after antigen exposure in guinea pig airways *in vitro*. *Am J Respir Crit Care Med* 156: 217–222

38 Ohrui T, Sekizawa K, Yamauchi K, Ohkawara Y, Nakazawa H, Aikawa T, Sasaki H, Takishima T (1991) Chemical oxidant potentiates electrically- and acetylcholine-induced contraction in rat trachea: possible involvement of cholinesterase inhibition. *J Pharmacol Exp Ther* 259: 371–376

39 Spicuzza L, Barnes PJ, Di Maria GU, Belvisi MG (2001) Effect of 8-iso-prostaglandin $F_{2\alpha}$ on acetylcholine release from paraxympathetic nerves in guinea pig airways. *Eur J Pharmacol* 416: 231–234
40 Adcock IM, Brown CR, Kwon OJ, Barnes PJ (1994) Oxidative stress induces NF-κB DNA binding and inducible NOS mRNA in human epithelial cells. *Biochem Biophys Res Commun* 199: 1518–1524
41 Rusznak C, Devalia JL, Sapsford RJ, Davies RJ (1996) Ozone-induced mediator release from human bronchial epithelial cells *in vitro* and the influence of nedocromil sodium. *Eur Respir J* 9: 2298–2305
42 Tomita K, Barnes PJ, Adcock IM (2003) The effect of oxidative stress on histone acetylation and IL-8 release. *Biochem Biophys Res Comm* 301: 572–577
43 Rahman I (2003) Oxidative stress, chromatin remodeling and gene transcription in inflammation and chronic lung diseases. *J Biochem Mol Biol* 36: 95–109
44 Pryor WA, Stone K (1993) Oxidants in cigarette smoke. Radicals, hydrogen peroxide, peroxynitrate, and peroxynitrite. *Ann NY Acad Sci* 686: 12–27
45 Britton JR, Pavord ID, Richards KA, Knox AJ, Wisniewski AF, Lewis SA, Tattersfield AE, Weiss ST (1995) Dietary antioxidant vitamin intake and lung function in the general population. *Am J Respir Crit Care Med* 151: 1383–1387
46 Schunemann HJ, Freudenheim JL, Grant BJ (2001) Epidemiologic evidence linking antioxidant vitamins to pulmonary function and airway obstruction. *Epidemiol Rev* 23: 248–267
47 Montuschi P, Barnes PJ (2002) Analysis of exhaled breath condensate for monitoring airway inflammation. *Trends Pharmacol Sci* 23: 232–237
48 Paredi P, Kharitonov SA, Barnes PJ (2002) Analysis of expired air for oxidation products. *Am J Respir Crit Care Med* 166: S31–S37
49 Dekhuijzen PNR, Aben KHH, Dekker I, Aarts LPHJ, Wielders PLM, van Herwarden CLA, Bast A (1996) Increased exhalation of hydrogen peroxide in patients with stable and unstable chronic obstructive pulmonary disease. *Am J Respir Crit Care Med* 154: 813–816
50 Nowak D, Kasielski M, Antczak A, Pietras T, Bialasiewicz P (1999) Increased content of thiobarbituric acid-reactive substances and hydrogen peroxide in the expired breath condensate of patients with stable chronic obstructive pulmonary disease: no significant effect of cigarette smoking. *Respir Med* 93: 389–396
51 Montuschi P, Collins JV, Ciabattoni G, Lazzeri N, Corradi M, Kharitonov SA, Barnes PJ (2000) Exhaled 8-isoprostane as an *in vivo* biomarker of lung oxidative stress in patients with COPD and healthy smokers. *Am J Respir Crit Care Med* 162: 1175–1177
52 Biernacki W, Kharitonov SA, Barnes PJ (2002) 8-isoproastane in exhaled condensate in patients with exacerbations of COPD. *Am J Respir Crit Care Med* 165: A447
53 Pratico D, Basili S, Vieri M, Cordova C, Violi F, Fitzgerald GA (1998) Chronic obstructive pulmonary disease is associated with an increase in urinary levels of isoprostane $F_{2\alpha}$-III, an index of oxidant stress. *Am J Respir Crit Care Med* 158: 1709–1714

54 Montuschi P, Kharitonov SA, Barnes PJ (2001) Exhaled carbon monoxide and nitric oxide in COPD. *Chest* 120: 496–501
55 Paredi P, Kharitonov SA, Leak D, Ward S, Cramer D, Barnes PJ (2000) Exhaled ethane, a marker of lipid peroxidation, is elevated in chronic obstructive pulmonary disease. *Am J Respir Crit Care Med* 162: 369–373
56 Maziak W, Loukides S, Culpitt S, Sullivan P, Kharitonov SA, Barnes PJ (1998) Exhaled nitric oxide in chronic obstructive pulmonary disease. *Am J Respir Crit Care Med* 157: 998–1002
57 Rutgers SR, Van der Mark TW, Coers W, Moshage H, Timens W, Kauffman HF, Koeter GH, Postma DS (1999) Markers of nitric oxide metabolism in sputum and exhaled air are not increased in chronic obstructive pulmonary disease. *Thorax* 54: 576–580
58 Agusti AG, Villaverde JM, Togores B, Bosch M (1999) Serial measurements of exhaled nitric oxide during exacerbations of chronic obstructive pulmonary disease. *Eur Respir J* 14: 523–528
59 Kharitonov SA, Robbins RA, Yates D, Keatings V, Barnes PJ (1995) Acute and chronic effects of cigarette smoking on exhaled nitric oxide. *Am J Resp Crit Care Med* 152: 609–612
60 Ichinose M, Sugiura H, Yamagata S, Koarai A, Shirato K (2000) Increase in reactive nitrogen species production in chronic obstructive pulmonary disease airways. *Am J Resp Crit Care Med* 160: 701–706
61 Corradi M, Montuschi P, Donnelly LE, Pesci A, Kharitonov SA, Barnes PJ (2001) Increased nitrosothiols in exhaled breath condensate in inflammatory airway diseases. *Am J Respir Crit Care Med* 163: 854–858
62 Rahman I, Morrison D, Donaldson K, Macnee W (1996) Systemic oxidative stress in asthma, COPD, and smokers. *Am J Respir Crit Care Med* 154: 1055–1060
63 Rahman I, van Schadewijk AA, Crowther AJ, Hiemstra PS, Stolk J, Macnee W, De Boer WI (2002) 4-Hydroxy-2-nonenal, a specific lipid peroxidation product, is elevated in lungs of patients with chronic obstructive pulmonary disease. *Am J Respir Crit Care Med* 166: 490–495
64 Di Stefano A, Caramori G, Gates T, Capelli A, Lusuardi M, Gnemmi I, Ioli F, Chung KF, Donner CF, Barnes PJ, Adcock IM (2002) Increased expression of NF-κB in bronchial biopsies from smokers and patients with COPD. *Eur Respir J* 20: 556–563
65 Caramori G, Romagnoli M, Casolari P, Bellettato C, Casoni G, Boschetto P, Fan Chung K, Barnes PJ, Adcock IM, Ciaccia A et al (2003) Nuclear localisation of p65 in sputum macrophages but not in sputum neutrophils during COPD exacerbations. *Thorax* 58: 348–351
66 Taggart C, Cervantes-Laurean D, Kim G, McElvaney NG, Wehr N, Moss J, Levine RL (2000) Oxidation of either methionine 351 or methionine 358 in α1-antitrypsin causes loss of anti-neutrophil elastase activity. *J Biol Chem* 275: 27258–27265
67 Vestbo J, Sorensen T, Lange P, Brix A, Torre P, Viskum K (1999) Long-term effect of inhaled budesonide in mild and moderate chronic obstructive pulmonary disease: a randomised controlled trial. *Lancet* 353: 1819–1823

68 Pauwels RA, Lofdahl CG, Laitinen LA, Schouten JP, Postma DS, Pride NB, Ohlsson SV (1999) Long-term treatment with inhaled budesonide in persons with mild chronic obstructive pulmonary disease who continue smoking. *N Engl J Med* 340: 1948–1953

69 Burge PS, Calverley PMA, Jones PW, Spencer S, Anderson JA, Maslen T (2000) Randomised, double-blind, placebo-controlled study of fluticasone propionate in patients with moderate to severe chronic obstructive pulmonary disease; the ISOLDE trial. *Br Med J* 320: 1297–1303

70 Lung Health Study Research Group (2000) Effect of inhaled triamcinolone on the decline in pulmonary function in chronic obstructive pulmonary disease. *N Engl J Med* 343: 1902–1909

71 Keatings VM, Jatakanon A, Worsdell YM, Barnes PJ (1997) Effects of inhaled and oral glucocorticoids on inflammatory indices in asthma and COPD. *Am J Respir Crit Care Med* 155: 542–548

72 Culpitt SV, Nightingale JA, Barnes PJ (1999) Effect of high dose inhaled steroid on cells, cytokines and proteases in induced sputum in chronic obstructive pulmonary disease. *Am J Respir Crit Care Med* 160: 1635–1639

73 Culpitt SV, Rogers DF, Shah P, de Matos C, Russell RE, Donnelly LE, Barnes PJ (2003) Impaired inhibition by dexamethasone of cytokine release by alveolar macrophages from patients with chronic obstructive pulmonary disease. *Am J Respir Crit Care Med* 167: 24–31

74 Hutchison KA, Matic G, Meshinchi S, Bresnick EH, Pratt WB (1991) Redox manipulation of DNA binding activity and BuGR epitope reactivity of the glucocorticoid receptor. *J Biol Chem* 266: 10505–10509

75 Okamoto K, Tanaka H, Ogawa H, Makino Y, Eguchi H, Hayashi S, Yoshikawa N, Poellinger L, Umesono K, Makino I (1999) Redox-dependent regulation of nuclear import of the glucocorticoid receptor. *J Biol Chem* 274: 10363–10371

76 Ito K, Barnes PJ, Adcock IM (2000) Glucocorticoid receptor recruitment of histone deacetylase 2 inhibits IL-1β-induced histone H4 acetylation on lysines 8 and 12. *Mol Cell Biol* 20: 6891–6903

77 Barnes PJ, Adcock IM (2003) How corticosteroid switch off inflammation in asthma. *Ann Int Med* 139: 359–370

78 Ito K, Lim S, Caramori G, Chung KF, Barnes PJ, Adcock IM (2001) Cigarette smoking reduces histone deacetylase 2 expression, enhances cytokine expression and inhibits glucocorticoid actions in alveolar macrophages. *FASEB J* 15: 1100–1102

79 Ito K, Watanabe S, Kharitonov S, Hanazawa T, Adcock IM, Barnes PJ (2001) Histone deacetylase activity and gene expression in COPD patients. *Eur Resp J* 18: 316S

80 Majo J, Ghezzo H, Cosio MG (2001) Lymphocyte population and apoptosis in the lungs of smokers and their relation to emphysema. *Eur Respir J* 17: 946–953

81 Kasahara Y, Tuder RM, Taraseviciene-Stewart L, Le Cras TD, Abman S, Hirth PK, Wattenberger J, Voelkel NF (2000) Inhibition of VEGF receptors causes lung cell apoptosis and emphysema. *J Clin Invest* 106: 1311–1319

82 Gotoh Y, Cooper JA (1998) Reactive oxygen species- and dimerization-induced activa-

tion of apoptosis signal-regulating kinase 1 in tumor necrosis factor-α signal transduction. *J Biol Chem* 273: 17477–17482
83 Barnes PJ (2001) New treatments for chronic obstructive pulmonary disease. *Curr Opin Pharmacol* 1: 217–222
84 Macnee W (2000) Oxidants/Antioxidants and COPD. *Chest* 117: 303S–317S
85 Grandjean EM, Berthet P, Ruffmann R, Leuenberger P (2000) Efficacy of oral long-term N-acetylcysteine in chronic bronchopulmonary disease: a meta-analysis of published double-blind, placebo-controlled clinical trials. *Clin Ther* 22: 209–221
86 Poole PJ, Black PN (2001) Oral mucolytic drugs for exacerbations of chronic obstructive pulmonary disease: systematic review. *Br Med J* 322: 1271–1274
87 Cuzzocrea S, Riley DP, Caputi AP, Salvemini D (2001) Antioxidant therapy: a new pharmacological approach in shock, inflammation, and ischemia/reperfusion injury. *Pharmacol Rev* 53: 135–159
88 Hansel TT, Kharitonov SA, Donnelly LE, Erin EM, Currie MG, Moore WM, Manning PT, Recker DP, Barnes PJ (2003) A selective inhibitor of inducible nitric oxide synthase inhibits exhaled breath nitric oxide. FASEB J 17: 1298–1300
89 Sies H, Masumoto H (1997) Ebselen as a glutathione peroxidase mimic and as a scavenger of peroxynitrite. *Adv Pharmacol* 38: 229–246

established emphysema [15]. More recent studies have nevertheless confirmed an increase in neutrophil proteins in patients even with subclinical emphysema detected by high resolution CT scanning [16]. The failure to detect the enzyme activity in lavage samples therefore may well represent the fact that these samples are a mixture of both the neutrophil products and naturally-occurring inhibitors that are present within airway secretions and hence not truly reflective of the environment in the interstitium leading to tissue damage and the development of emphysema (see later).

Pathological studies have shown that the amount of neutrophil elastase detected immunologically in the interstitium does relate to the severity of emphysema [17]. In addition, there is clearly a relationship between the amount of mucus production and the concentration of active neutrophil elastase in the secretions [14]. More recent studies have shown that activated neutrophils result in mucus secretion that is dependent on adhesion *via* the CD11b/18 surface receptor and elastase on the surface of the neutrophil [18]. Although the exact mechanism has yet to be determined it seems likely that the epidermal growth factor receptor on mucous glands plays a key role leading to further mucus production [19].

Studies of airway secretions from patients with chronic bronchitis indicate that there is a toxic component that influences ciliary beating, and this can be inhibited by specific inhibitors of neutrophil elastase [13], suggesting the enzyme plays a role. Furthermore, when neutrophil elastase has been detected in lavage samples from patients with established emphysema, cessation of smoking leads to a reduction in this enzyme activity [20]. As cessation of smoking is known to be the only intervention that prevents progression of emphysema, this positive relationship between the smoking habit and a potential pathogenic enzyme adds support to its role. Finally, in the airway secretions neutrophil elastase is readily detected when the secretions are purulent (i.e., when there is a major neutrophil influx). This feature of secretions is related to bacterial colonisation [21] and in particular to bacterial exacerbations of COPD [22]. Such episodes have been shown to relate to progressive loss of lung function [23, 24] which at least implicates this enzyme in damage at bronchial, if not at alveolar, level.

Factors involved in elastase delivery to the lung

As the enzyme is predominantly contained within circulating neutrophils, it is delivered to the lung as part of the inflammatory process. This involves the activation, endothelial adhesion, migration and degranulation of the neutrophil. Specific neutrophil chemoattractants (such as interleukin (IL)-8 and leukotriene (LT) B4) have been identified in lung lavage fluids [25–27] and bronchial secretions [28] in patients with established COPD and emphysema. In addition, endothelial adhesion molecules have been shown to be up-regulated on pathological specimens and relate

Table 1 - Effects of neutrophil elastase in vitro and in vivo

Proteolysis	
Degradation	Elastin, fibronectin and other matrix components (for review see [103])
	Immunoglobulin A [104]
	T lymphocyte surface antigens [105]
	IL-6 [106]
	Cystatin C [70], TIMPs [83, 107]
Activation	Cathepsin B [70]
	TGF-β [108]
	MMP-2 [85,109], MMP-3 [85], MMP-9 [84]
	Complement components [110, 111]
Cellular	
Neutrophils and monocytes	Complexed with α_1-antitrypsin (α_1-AT), chemotactic for neutrophils [112]
	In complex with α_1-AT, increased α_1-AT secretion by monocytes and alveolar macrophages [113] (and hepatoma cell line [114])
	Increased LTB4 release by macrophages [27]
Epithelium	Disruption and detachment [12]
	Enhanced production of prostaglandin E2 [115] and IL-8 [116]
	Reduced production of SLPI [117, 118] and elafin [119]
	Increased mucin MUC5AC protein content [120]
Endothelium	Induced apoptosis [121]
Bacteria	Killing effect in vitro [122]
	Increased likelihood of adherence and colonization [123–125]
Genetic	
Transcription	Increased α_1-AT expression by monocytes and alveolar macrophages [113]
	Increased SLPI [117, 118, 126] and elafin [119] expression (high concentrations of neutrophil elastase only)
	Increased expression of MUC5AC [120] and IL-8 [127]

is evidence therefore to implicate neutrophil elastase in most of these pathological processes and clinical features [14].

However neutrophil elastase activity was difficult to identify in early lavage studies from patients with emphysema, although there has always been a general acceptance that neutrophil numbers are increased both in smokers and patients with

Neutrophil elastase

Neutrophil elastase is a 30 kDa enzyme that is predominantly localized to the polymorphonuclear leukocyte, although it is also found within a subset of monocytes [1]. The enzyme is produced during neutrophil maturation, and the gene is only expressed early during this process, as the cell matures from a promyelocyte to a metamyelocyte [2]. Thereafter gene transcription ceases and the enzyme is packaged into the primary azurophil granule and is released in the active form only on cell activation and degranulation. The enzyme itself is made as a pre-proenzyme, which is cleaved at both the carboxy and amino terminal ends, releasing the mature enzyme which is stored in its fully active form [3]. The enzyme is a classical serine proteinase with serine at its active site and a catalytic triangle of serine, histidine and asparagine [4] that gives it its specificity for both enzyme activity and inhibition by inhibitors.

The enzyme was named neutrophil elastase because of the cell in which it is stored and its ability to digest the connective tissue elastin. Nevertheless, despite this function the enzyme is also able to cleave a variety of other substrates including the connective tissue components fibronectin, laminin and collagen, cell surface receptors (see Tab. 1) and fibrinogen, potentially leading to intravascular coagulation.

The role of neutrophil elastase in chronic lung disease was explored following the early identification of severe early onset emphysema in patients with the inherited deficiency of α_1-antitrypsin (α_1-AT, see later). Since α_1-AT was the major serum serine proteinase inhibitor, studies were undertaken to identify a serine proteinase that could produce emphysema. Early animal studies demonstrated that enzymes such as the plant enzyme papain and subsequently pancreatic porcine elastase (another serine proteinase) were able to induce pathological changes in experimental animals that resembled human emphysema ([5] and [6] respectively). In view of these observations, studies were undertaken to identify a human enzyme with the same properties. Neutrophil extracts were shown to induce emphysema [7] and subsequently neutrophil elastase was purified and confirmed to have the same ability [8]. This led to a general acceptance that neutrophil elastase was probably the major enzyme that resulted in the development of human emphysema and there is now an extensive literature involving both *in vitro* and *in vivo* data from experimental animal and human studies exploring the mechanisms involved. Indeed, although much of the early work was confined to the study of emphysema, proteolytic enzymes such as neutrophil elastase have now been implicated in the more generic disease grouping COPD, where emphysema may be only one component or even absent. This is largely related to experimental data that confirms neutrophil elastase has the ability to produce dramatic bronchial damage [9], to lead to mucous gland hyperplasia [10] and mucus secretion [11] as well as damage to bronchial epithelium [12] and a reduction in ciliary beat frequency of epithelial cells [13]. These are all clinical features of subsets of patients encompassed in the global term COPD and there

Proteinases in COPD

Anita L. Sullivan and Robert A. Stockley

Department of Respiratory Medicine, 1st Floor Nuffield House, Queen Elizabeth Hospital, Edgbaston, Birmingham B15 2TH, UK

Introduction

Proteolytic enzymes play a key role in a multitude of diverse normal physiological processes. These range from digestion of food to the intracellular degradation of antigens or subtle post-translational processing of proteins. In addition, proteolytic enzymes can play a key role in the initiation or resolution of cascade processes such as coagulation and the complement cascade.

The importance of these enzymes only becomes apparent when they are deficient, as in pancreatic disease leading to malabsorption, or overactive as with C1 esterase in the generation of angio-oedema. However, for their normal physiological function, processes must be in place to modulate their activity. In particular the enzymes need to be "switched on" when required and "switched off" when no longer needed or limited in their radius of activity to prevent excessive collateral damage.

This concept of collateral damage has become of major importance in chronic lung disease where enzymes have been implicated in the destructive processes that lead to the pathological and clinical features of the disease. This has been polarised into the proteinase-antiproteinase theory of chronic lung disease that has dominated research initially in emphysema and more recently in the broader spectrum of diseases termed chronic obstructive pulmonary disease (COPD). The theory predicts that, in the presence of inflammation, normal proteinase function is retained while collateral damage is prevented by specific proteinase inhibitors. However when the enzyme is released in excess amounts or the quantity or function of the inhibitors is reduced, the resultant excessive enzyme activity leads to the tissue damage that is central to the pathogenesis of the lung disease.

The current chapter explores the mechanisms involved with particular reference to neutrophil elastase.

to the number of neutrophils [29]. Neutrophils from patients with COPD and emphysema show increased expression of adhesion molecules [30] as well as an enhanced ability to migrate and degranulate to an appropriate signal [31]. The greater or more continuous this process, the more likely that neutrophil mediated lung damage is to occur and this will be specifically important in patients who have deficiency of controlling neutrophil elastase inhibitors such as α_1-AT.

Neutrophil recruitment also has the ability to amplify the inflammatory process especially in α_1-AT deficiency. Early studies of such patients identified the neutrophil chemoattractant LTB4 in bronchoalveolar lavage samples [27]. It was shown that macrophages from such patients were the source of LTB4 and that neutrophil elastase had the ability to activate macrophages to produce even more LTB4. These observations led to the concept that release of elastase by neutrophils recruited to the lung could amplify the process because deficiency of α_1-AT failed to inactivate the enzyme *in situ*. This concept is outlined in Figure 1. In addition, neutrophil elastase has been shown to activate epithelial cells leading to the release of a further neutrophil chemoattractant IL-8, again a process that would potentially be greater in subjects with α_1-AT deficiency. Indeed, studies of bronchial secretions from patients with established COPD with emphysema have shown more inflammation and greater concentrations of at least LTB4 in subjects with α_1-AT deficiency [32] supporting these concepts.

However the process may be even more complicated since activated neutrophils can release both LTB4 [33] and IL-8 [34] thereby amplifying the inflammatory process directly. This may be particularly true when the airways are colonised by bacteria and particularly during exacerbations due to bacterial infections. During such episodes, even in patients with normal α_1-AT, free neutrophil elastase activity is easily detectable within the airway secretions [22] and hence has the potential to at least cause the pathological changes at bronchial level seen in COPD while also amplifying the inflammatory process as indicated above.

Where is the enzyme effective?

Neutrophil elastase is stored within the azurophil granules and when the neutrophil is activated, these granules are exocytosed and the enzyme is released. Studies have shown that the concentration of enzyme within the azurophil granule is approximately 5 mM [35] and as the enzyme is released from the granule, it diffuses away and its concentration becomes reduced in an exponential manner (Fig. 2). This process is of major importance in understanding how tissue damage can occur. The most active inhibitor of neutrophil elastase is α_1-AT. In normal subjects the serum concentration during health is approximately 30 µM i.e., approximately two orders of magnitude lower in concentration than the enzyme within the azurophil granule. Since α_1-AT inhibits neutrophil elastase on a one to one molar basis, the high con-

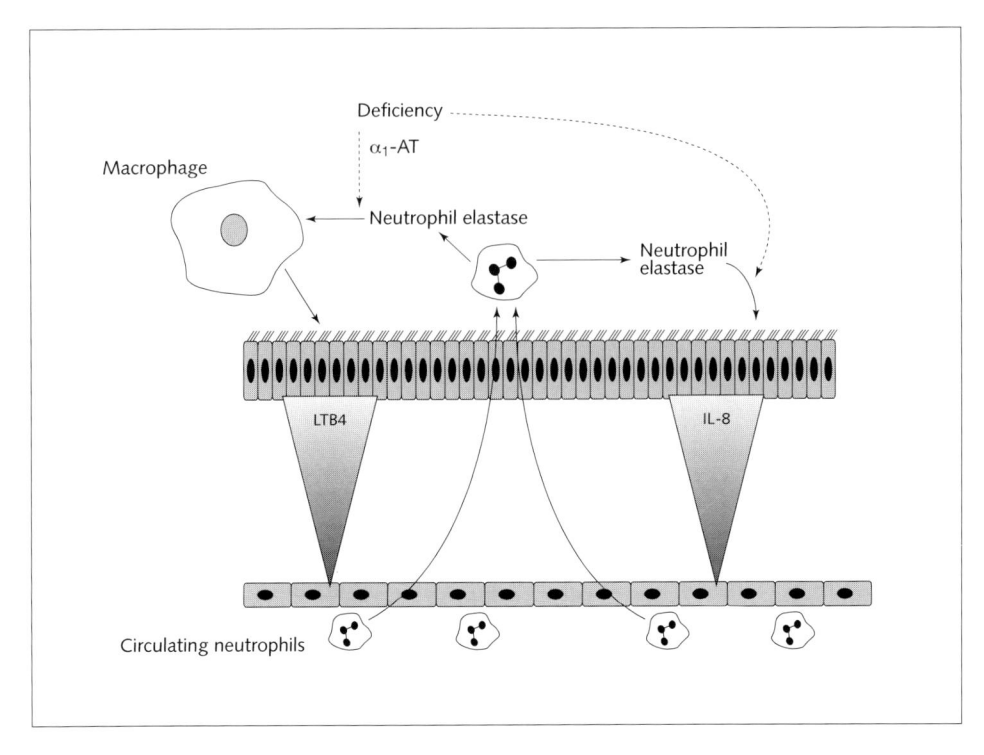

Figure 1
The process hypothesized by Hubbard et al. [27] to explain the amplification of neutrophil recruitment in patients with α_1-antitrypsin (α_1-AT) deficiency. Activated macrophages release LTB4 which attracts circulating neutrophils. Neutrophils release elastase, which would normally be inactivated by α_1-AT. However in deficiency the elastase persists and triggers more LTB4 release from macrophages and stimulates release of IL-8 from epithelial cells, thereby amplifying neutrophil influx.

centration of enzyme as it is released from the granule means that it cannot be inhibited completely until its concentration has fallen to that of its physiological inhibitors. Thus, there is always an area of obligate enzyme activity around a released granule, which can cause tissue damage in the vicinity of an activated neutrophil even in patients with normal antiproteinases.

From the shape of the predicted exponential curve, it can be seen that the radius of low enzyme concentration increases exponentially as the concentration drops below 10 μmol (Fig. 3). Patients with α_1-AT deficiency have a serum concentration of inhibitor of approximately 5 μM. Inhibitor concentrations as low as this would result in a much larger area of persistent enzyme activity around the released granule in such subjects until the concentration equals that of α_1-AT. This concept has

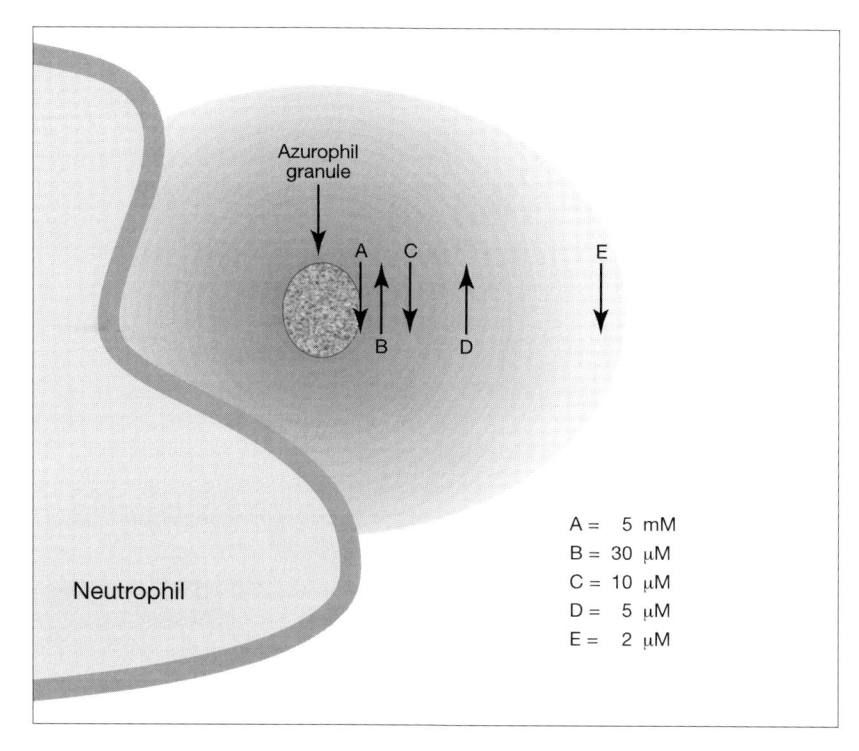

Figure 2
Diagrammatic representation of the change in neutrophil elastase concentration as it is released from an azurophil granule after exocytosis from the neutrophil. Adjacent to the granule the concentration is high, but it decreases exponentially; the arrows represent concentrations at given distances from the granule. In patients with normal α_1-AT concentration, the enzyme activity would be completely blocked at B (30 µM), whereas α_1-AT deficiency, the enzyme activity would not be inhibited until it has diffused far enough away for the concentration to fall to 5 µM.

been supported by experimental data utilising both purified inhibitors [35] and plasma from patients with different inherited α_1-AT concentrations [36]. These observations are thought to explain, at least in part, the reason that patients with α_1-AT deficiency develop rapidly progressive extensive emphysema.

In addition to the quantum proteolysis theory outlined above, studies have shown that enzymes released during the exocytosis of the azurophil granules can also spread and remain on the cell surface [37]. At this site, the enzymes become more resistant to inactivation and may play a key role in tissue damage as the neutrophils adhere to their substrate. Whether cell surface proteinases are important in

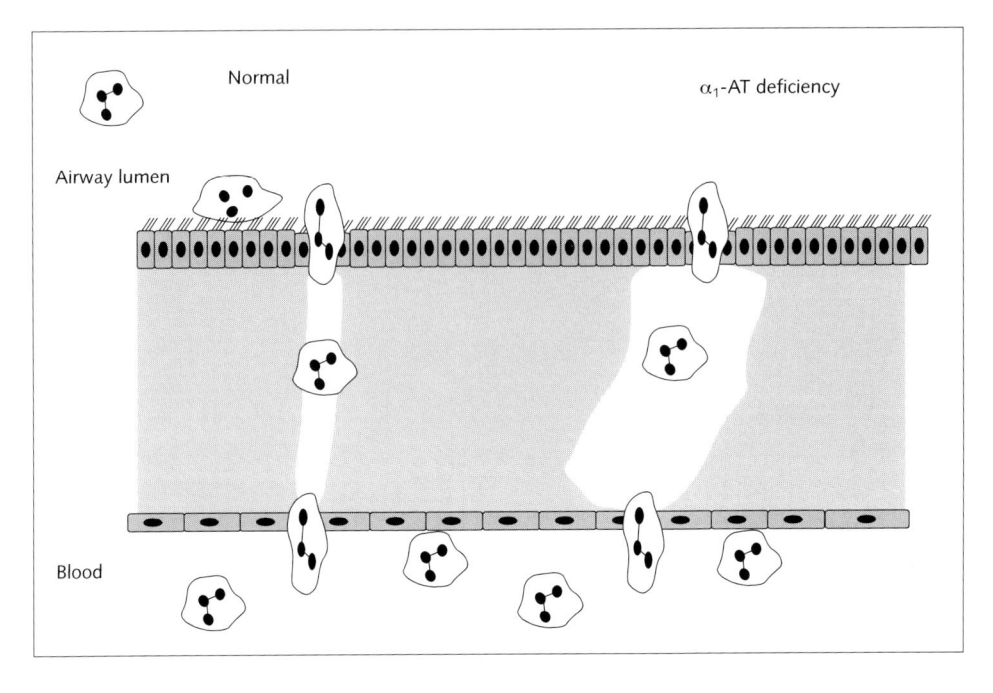

Figure 3
Diagrammatic representation of the effects of the varying radius of enzyme activity from azurophil granules released as neutrophils migrate from circulation to airway lumen. In patients with α_1-AT deficiency, the radius of activity is much greater and hence more tissue is destroyed during neutrophil migration.

normal physiological processes for the neutrophil remains uncertain. However, studies have shown that specific inhibitors of neutrophil elastase can down-regulate neutrophil responses such as migration, even using experimental studies that do not involve connective tissue substrates [38]. It is possible that this has something to do with the processes of adhesion and activation of adhesion molecules, although clearly further studies need to be carried out.

However, in support of this concept, neutrophils from patients with Chediak-Higashi syndrome fail to respond to chemoattractants [39]. This disease is a rare and usually fatal condition where a defect in the *CHS1/LYST* gene, coding for a protein whose function is unknown, leads to the development of giant lysosomal structures within cells, particularly haematopoietic and neurological tissue [40]. In neutrophils, only the azurophil granules are affected, becoming giant cytoplasmic granules [41]. In studies on peripheral blood neutrophils and bone marrow from affected

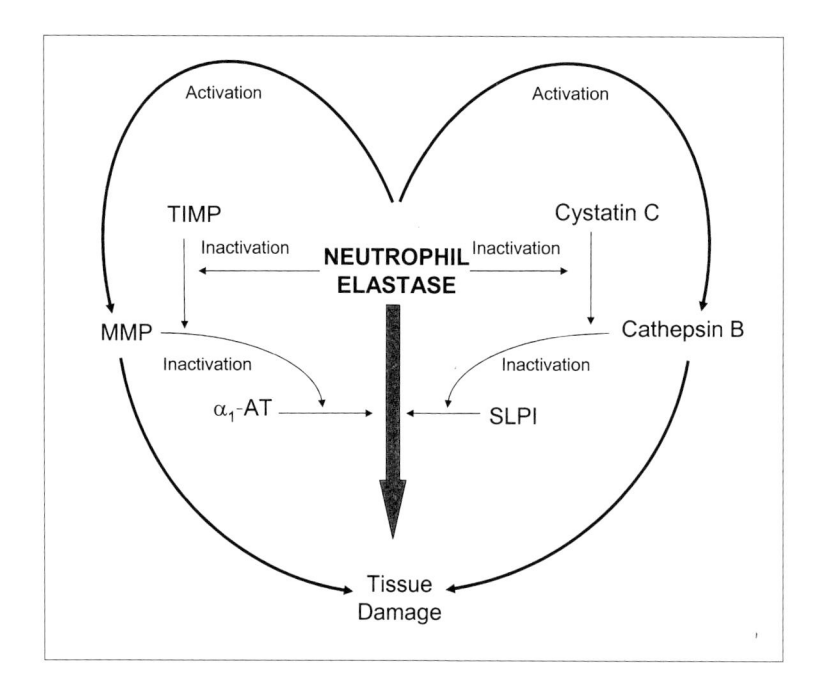

Figure 4
The proteinase-antiproteinase cascade is at the "heart" of the process. Neutrophil elastase (NE) is prevented from causing tissue damage by α_1-AT and SLPI. α_1-AT can be inactivated by MMPs and this can be facilitated by NE as it will both activate MMPs by cleavage of proenzymes or inactivate TIMPs. Similarly SLPI can be inactivated by cathepsin B and this can also be facilitated by NE through activation of pro cathepsin B and inactivation of cystatin C. Both MMPs and cathepsin B can also cause tissue damage directly.

patients, marrow cells contained normal quantities of neutrophil elastase and cathepsin G mRNA, but reduced enzyme concentrations (as measured by enzyme-linked immunoabsorbent assay and by activity assays), while mature circulating neutrophils contained no measurable enzyme activity and the azurophil granules were negative on immunohistochemical staining for neutrophil elastase and cathepsin G [42], suggesting that the defect lies in the movement of enzymes to the azurophil granules rather than in transcription or activation. In another study, neutrophils from human patients with Chediak-Higashi syndrome exhibited markedly reduced exocytosis of myeloperoxidase from giant granules although that from secondary and specific granules was relatively normal [41]. An animal model with an analogous genetic defect to Chediak-Higashi syndrome exists, namely the beige mouse. These mice fail to have an appropriate neutrophilic response to acute infections, although interestingly this is associated with greater survival suggesting that a neutrophilic response and proteolytic enzyme release may play some role in mortality [43]. Studies on neutrophils from beige mice, however, have shown that they produce normal levels of cathepsin G and a latent proform of neutrophil elastase that can be activated after release from the cell [44]. Furthermore a recent study found that it was possible to induce a neutrophilic response and emphysema in these animals when exposed to intratracheal formyl-methionyl-leucyl-phenylalanine [45] suggesting the beige mouse is not entirely synonymous with Chediak-Higashi syndrome.

Another inherited neutrophil disorder, the Papillon-Lefèvre syndrome, is the result of a defect in the gene for cathepsin C (also known as dipeptidyl peptidase I), a lysosomal protease that activates the azurophil granule proteinases by cleaving a prodipeptide from the N-terminus. These patients suffer severe destructive periodontitis in early life, as well as palmoplantar keratosis [46], and an animal model has been developed [47]. While most studies have shown reduced random and directed migration by neutrophils, impaired fusion of phagosomes with lysosomes and impaired bacterial killing by neutrophils, the lack of functional neutrophil serine proteinases does not appear to cause systemic immunodeficiency or impair survival, suggesting there may be a degree of 'redundancy' in the function of these enzymes, or else that they can be activated by other mechanisms after secretion as proenzymes.

The role of proteolytic enzymes in the process of neutrophil migration from blood vessel to airway lumen is still uncertain. Although migration of neutrophils from patients with Chediak-Higashi syndrome [39] and Papillon-Lefèvre syndrome [48] is impaired, neutrophils from neutrophil elastase (NE) and cathepsin G (CG) knockout mice migrate normally *in vitro* to chemotactic stimuli (NE [49]; CG [50]) and *in vivo* (NE [51, 52]; CG [50]) in response to infection, while bacterial killing and survival are adversely affected only in neutrophil elastase knockout mice [50]. This may also relate to redundancy of the various azurophil granule proteinases not only in knockout mice but also in humans, since the activity of all the azurophil granule proteinases is affected in Chediak-Higashi and Papillon-Lefèvre syndrome. Even double knockout mice however, lacking both neutrophil elastase and cathepsin G, are able to mount a neutrophilic response with infiltration of similar or greater numbers of neutrophils into tissues both in response to fungal infection and in a lipopolysaccharide-induced shock model, although clearance of fungi was impaired in single knockout mice and severely impaired in double knockout mice [51]. This again suggests the enzymes are not critical for cell migration. The fact that proteinase inhibitors fail to inhibit neutrophil chemotaxis across endothelial basement membranes in some *in vitro* studies [53–55] also supports this concept. However studies by other groups have suggested the enzymes do play a role, at least in migration across acellular membranes [38, 56, 57]. The issue therefore remains unresolved at present. Interestingly, double knockout mice were also protected from the diffuse alveolar damage, vascular leak and shock lung that developed in the wild-type mice in response to endotoxin [51] suggesting that these serine proteinases are important effectors in tissue damage.

A list of the actions of neutrophil elastase possibly relevant to the pathological process in COPD is given in Table 1.

Other enzymes

Proteolytic enzymes other than neutrophil elastase have also been studied, though in less detail even though several have the ability to generate features of chronic lung disease. Nevertheless the basic principles that govern their role will be similar to those for neutrophil elastase.

Other neutrophil serine proteinases

Neutrophil cathepsin G (also stored within the azurophil granule) has been shown to induce mucous gland hyperplasia *in vivo* [9] and mucus secretion *in vitro* [11]. The third elastinolytic enzyme stored within the azurophil granule is proteinase 3. Again this enzyme has been shown to induce mucus secretion [58], but in addition has been shown to generate emphysema in experimental animals [59]. The major function of this enzyme remains unknown although it is expressed on the surface of neutrophils and is one of the two antigens for the autoantibodies associated with Wegener's granulomatosis [60]. Indeed activation of neutrophils by cross-linking surface proteinase 3 and the FcγR11A receptor by autoantibodies is thought to be central to the pathogenesis of the disease [61]. This process of cross-linking can be abrogated by α_1-AT [62], which may explain why Wegener's granulomatosis appears to be both more common [63] and more aggressive [64] in patients with α_1-AT deficiency.

Other serine proteinases identified in the airways include mast cell-derived tryptase and chymotryptase, which can degrade matrix components, activate proteinase-activated receptors and matrix metalloproteinases, and help to recruit inflammatory cells [65]; their significance however in COPD is unknown.

Cysteine proteinases

The cysteine proteinase cathepsin B has been shown to induce emphysema [66] and bronchial disease [67] in experimental animals. The source of this enzyme is largely unknown, although it may be produced by macrophages [68] and epithelial cells [69]. The enzyme however is secreted in a latent form into the airways and has to be activated by cleavage of the pro-enzyme, which can be achieved with neutrophil elastase [70]. It is possible therefore that if cathepsin B plays a role in lung disease it is *via* a cascade that requires neutrophil elastase in order to activate the cathepsin

B. Neutrophil elastase can also inactivate cystatin C, the main inhibitor of cathepsin B [70], while cathepsins B, L and S can inactivate secretory leukoproteinase inhibitor (SLPI) a major physiological inhibitor of NE in the lungs [71], and cathepsin L can inactivate α_1-AT [72] (see Fig. 3). Cathepsins L and S are produced by macrophages, and have elastinolytic activity [73, 74], but their role in COPD is not clear. In lung secretions, more cystatin C and cathepsin L are present in bronchoalveolar lavage fluid (BALF) from smokers with emphysema compared to smokers without emphysema [75]. Whether this indicates a pathogenic proteinase/antiproteinase balance remains unknown.

Metalloproteinases

These enzymes are named for their proteolytic activity on matrix constituents, requiring a metal ion within their structure for catalytic activity. They constitute a large family of structurally similar enzymes, present throughout the body, with an important role in cell signalling as well as in matrix remodelling. Nomenclature is complex and a reference list is given in Table 2.

Recent studies have suggested that metalloproteinases may play a role in the pathogenesis of emphysema. Macrophages and neutrophils produce a number of metalloproteinase elastases and collagenases. Studies with knockout mice for macrophage metalloelastase (MMP-12) have indicated that they are fully protected against the development of emphysema in response to cigarette smoke [76], whilst knocking out the neutrophil elastase gene in mice only leads to reduction of the development of cigarette-smoke induced emphysema by a third [77]. It is possible however that in MMP-12 knockout mice, protection against emphysema was related to failure to induce a neutrophilic inflammatory response rather than loss of the direct elastinolytic effect of MMP-12 itself. Further studies have shown that the extent of acute smoke-induced elastin and collagen degradation, as measured by breakdown products in BALF, correlated with the number of neutrophils, not macrophages [78]. Furthermore, in a guinea-pig model of cigarette smoke-induced emphysema, administration of a synthetic neutrophil elastase inhibitor reduced the neutrophil recruitment into the lungs and the acute elastin and collagen degradation to control levels, and significantly reduced airspace enlargement [79]. In mice exposed to cigarette smoke, α_1-AT reduced cytokine transcription and release, suggesting that the inflammation was also serine proteinase dependent [80]. Other studies have shown that α_1-AT (which will not inhibit metalloproteinase activity) can prevent the early connective tissue destruction that occurs when these animals are exposed to smoke, indicating that serine proteinases play at least an early role [78]. Furthermore, although MMP-12 is the predominant metalloelastase of the murine macrophage, human macrophages produce a broader spectrum of elastases. Therefore, while these animal models clearly demonstrate the importance of MMP-

Table 2 - Matrix metalloproteinases studied in COPD

MMP	Name	Main ECM substrates	Serine protease inhibitors degraded*	Other substrates
MMP-1	Collagenase-1	Collagens I–III, VII, VIII, X, XI, fibronectin, gelatin, laminin	α_1-ACT, α_2-M, α_1-AT	fibrin, fibrinogen, IL-1β, proTNF-α, C1q
MMP-2	Gelatinase-A	Collagens I, III–V, VII, X, XI, elastin, fibrillin, fibronectin, gelatin, laminin	α_1-ACT, α_1-AT	fibrin, fibrinogen, C1q, substance P, IL-1β, proTNF-α, proTGF-β, T kininogen
MMP-3	Stromelysin-1	Collagens III–V, VII, IX–XI, elastin, fibrillin, fibronectin, gelatin, laminin	α_1-ACT, α_2-M, α_1-AT	fibrin, fibrinogen, C1q, proTNF-α, IL-1β, E-cadherin, plasminogen
MMP-7	Matrilysin	Collagens I, IV, elastin, fibronectin, gelatin, laminin	α_1-AT	E-cadherin, proTNF-α, plasminogen, fibrinogen
MMP-8	Collagenase-2, neutrophil collagenase	Collagens I–III	α_2-M, α_1-AT	C1q, fibrinogen, substance P
MMP-9	Gelatinase B	Collagens IV, V, XI, XIV, elastin, fibrillin, gelatin, laminin	α_2-M, α_1-AT	C1q, fibrin and fibrinogen, IL-1β, proTGF-β and proTNF-α, plasminogen, substance P
MMP-10	Stromelysin-2	Collagens III–V, elastin, fibronectin, gelatin		fibrinogen
MMP-12	Macrophage metallo-elastase	Collagens I, IV, elastin, fibrillin, fibronectin, gelatin, laminin	α_2-M, α_1-AT	factor XII, fibrinogen, proTNF-α, plasminogen
MMP-14	Membrane type (MT)1-MMP	Collagens I-III, fibronectin, gelatin, laminin	α_2-M, α_1-AT	factor XII, fibrin, fibrinogen, proTNF-α, ProMMP2

*MMPs lose catalytic activity on binding α_2-macroglobulin because of steric hindrance, and the complexed MMP-α_2-macroglobulin is eventually cleared by endocytosis. α_1-ACT, α_1-antichymotrypsin; α_1-M, α_2-macroglobulin; α_1-AT, α_1-antitrypsin. Adapted from [128].

12 in the pathogenic processes involved in emphysema, the evidence still points to neutrophil elastase as a key effector enzyme in lung destruction in both mouse and man.

It is possible that cigarette smoking induces a cascade of events, which may involve inactivation of several inhibitors and activation of several enzymes. For instance, MMP-12 [81] and MMP-9 [82] have been shown to inactivate α_1-AT and neutrophil elastase inactivates the tissue inhibitors of metalloproteinases (TIMPs) [83] which are the natural inhibitors of enzymes such as MMP-12. Furthermore neutrophil elastase can activate MMP pro-enzymes [84, 85] suggesting an even more complex interaction, as illustrated in Figure 4. Either or both processes therefore would result in a failure to control the relevant proteinase appropriately, resulting in persistent activity of both the neutrophil and macrophage enzymes. It is likely that the only way to resolve the issues of the key enzyme effecting tissue damage in what could be a circuitous process will be the subsequent use of specific enzyme inhibitors in human studies.

More recently the role of TNF-α in acute lung tissue breakdown after cigarette smoke exposure in MMP-12 knockout mice has been assessed [86]. Smoke-exposed macrophages from MMP-12 knockout mice exhibited normal induction of expression of TNF-α mRNA but failed to release biologically active protein, and this could be reproduced with the use of an inhibitor of MMP-12 in macrophages from wild-type mice. The authors hypothesised that MMP-12 could convert latent TNF-α protein to an active form. Because TNF-α induces expression of endothelial adhesion molecules necessary for neutrophil recruitment, the lack of MMP-12 would prevent this response. The implications are that neutrophil elastase remains the key effector of lung damage while MMP-12 is crucial in initiating the inflammatory response by means of its ability to modulate adhesion molecule expression *via* TNF-α activation.

MMPs with collagenolytic activity have also been implicated in experimental animal models. A transgenic mouse expressing MMP-1 (human collagenase-1) had increased airspace enlargement suggestive of emphysema [87]. However this human enzyme would be expressed during neonatal and postnatal development, indicating that the pathological changes could be developmental rather than acquired. Nevertheless, there have been studies in man indicating that collagenase activity is detectable in lavage fluids and is probably derived from activated macrophages [88]. Again interpretation of these data is difficult with particular reference to cause or effect. Of interest, the patients with emphysema who had ceased smoking still had similar detectable collagenase activity [88]. Since smoking cessation is the single intervention that is proven to reduce the progression of emphysema, the continued presence of collagenase suggests it is an effect rather than cause of emphysema. Interestingly, in the same study neutrophil elastase activity was also detected in lavage samples but was largely absent in those individuals who had ceased smoking, providing further indirect evidence suggesting that neutrophil elastase plays a more important role.

Several studies have assessed the presence of MMPs in tissue and secretions from patients with COPD. Immunohistochemistry comparing lung tissue from patients with COPD and healthy controls have produced varying data. Increased tissue staining for MMP-1, MMP-8 (collagenase-2,), MMP-2 (gelatinase A) and MMP-9 (gelatinase B) was found in patients compared to controls [89]. On the other hand, in another study negative tissue staining for MMPs-1, -8, -9 and TIMP-1 protein was observed in lung tissue from both patients or controls although positive staining was seen for MT (membrane type) 1-MMP and MMP-2 protein in alveolar macrophages, pneumocytes and fibroblasts, and positive staining for MMPs-1, -2, -9 and TIMPs 1 and 2 mRNA. A statistically significant increase in MMP-2 and MMP-9 mRNA was found in emphysema. Furthermore, homogenized lung from patients with emphysema contained greater collagenolytic and gelatinolytic activity than controls, and the gelatinolytic activity originated from MMPs-2 and 9. MMP-12 elastinolytic activity was not found, nor was activity of NE significantly elevated in patients [90]. Positive staining for MMP-1 mRNA and protein in the type II pneumocyte and lung tissue has also been described in patients with COPD [91].

Studies on induced sputum showed increased MMP-9 and TIMP-1 in patients with COPD and asthma compared to controls, together with an increased cell count in COPD [92]. Furthermore the MMP-9 concentration correlated with the absolute numbers of neutrophils and macrophages. However the MMP-9 appeared to be in its inactive form, perhaps owing to the large excess of TIMP-1 in the stable state. A later study showed that the increased MMP-9 in sputum of patients with COPD and asthma was not due to increased secretion by blood granulocytes [93]. When macrophages were isolated from BALF from patients with COPD or controls, enhanced MMP-1 and MMP-9 expression was found, compared to macrophages from controls [94, 95]. Finlay et al. [88] found elevated mRNA and protease activity for MMPs-1 and 9 in AM from BALF from patients compared to controls, whereas expression of MMPs-2 and -12 was not altered, and secretion of these two enzymes did not appear to occur *in vitro*.

The variability in the findings of such observational studies in humans suggests that there may be changing profiles of MMP expression and activation throughout the natural history of the disease, in relation to current smoking or cessation of smoking. It should be noted however that patient characterization is usually superficial and the COPD phenotypes as well as source of the samples may vary greatly. Furthermore, it is necessary to assess enzyme activity as well as immunological presence, in view of the complex activation systems of MMPs. Finally MMPs often remain on the cell membrane where they participate in cell signaling activities, so that studies of secretions may not reveal the true contribution from these enzymes.

Nevertheless genetic polymorphisms of MMP-1, -9, -12 and TIMP-2 have been linked to smoking-induced emphysema [96–98] although the clinical and pathological significance remains uncertain at present.

Other roles of proteinases

Many allergens have been shown to be proteinases, which has raised interest in the mechanisms whereby these enzymes stimulate the immune system. It has been suggested that the enzymes digest intracellular adhesion molecules allowing them access to the sub-epithelial regions of the lung where they can directly stimulate the host immune response [99]. There is also some evidence to suggest that house dust mite proteinases may damage the antiproteinase screen [100] and components of the immune system such as CD23 [101]. Whether these enzymes play a role in COPD remains unknown.

Bacteria also release proteolytic enzymes, although the role of these enzymes is also largely unexplored. Nevertheless, recent information has suggested that a serine proteinase from *Haemophilus influenzae* may play a key role in promoting colony formation in the airway and therefore may be central to colonisation and infection [102]. This process could be central to airways inflammation and acute exacerbations of COPD. It remains to be seen whether this enzyme plays a role in patients with established lung disease and whether it is influenced by lung inhibitors.

Conclusion

In conclusion, proteinases potentially play a key role in the pathogenesis of many of the pathological and clinical features of COPD. Many enzymes have been implicated although lack of understanding of the limitations of animal models, diversity of methods of human sample collection and patient demography has led to some confusion. Understanding the mechanisms involved enables the rational development of new therapeutic strategies and these will be necessary to clarify the exact role of individual enzymes.

Acknowledgements
We would like to thank Dr. Anita Pye for assistance in the preparation of the figures.

References

1 Owen CA, Campbell MA, Boukedes SS, Stockley RA, Campbell EJ (1994) A discrete subpopulation of human monocytes expresses a neutrophil-like proinflammatory (P) phenotype. *Am J Physiol* 267: L775–L785
2 Fouret P, du Bois RM, Bernaudin JF, Takahashi H, Ferrans VJ, Crystal RG (1989) Expression of the neutrophil elastase gene during human bone marrow cell differentiation. *J Exp Med* 169: 833–845

3   Gullberg U, Bengtsson N, Bülow F, Garwicz D, Lindmark A, Olsson I (1999) Processing and targeting of granule proteins in human neutrophils. *J Immunol Methods* 232: 201–210
4   Dodson G, Wlodawer A (1998) Catalytic triads and their relatives. *Trends Biochem Sci* 23: 347–352
5   Gross P, Pfizer EH, Tolker B, Babyok MA, Kaschak M (1964) Experimental emphysema: its production with papain in normal and silicotic rats. *Arch Environ Health* 11: 50–58
6   Hayes JA, Korthy A, Snider GL (1975) The pathology of elastase-induced panacinar emphysema in hamsters. *J Pathol* 117: 1–14
7   Marco V, Mass B, Meranze DR, Weinbaum G, Kimbel P (1971) Induction of experimental emphysema in dogs using leukocyte homogenates. *Am Rev Respir Dis* 104: 595–598
8   Janoff A, Sloan B, Weinbaum G, Damiano V, Sandhaus RA, Elias J, Kimbel P (1977) Experimental emphysema induced with purified human neutrophil elastase: tissue localization of the instilled protease. *Am Rev Respir Dis* 115: 461–478
9   Lucey EC, Stone PJ, Breuer R, Christensen TG, Calore JD, Catanese A, Franzblau C, Snider GL (1985) Effect of combined human neutrophil cathepsin G and elastase on induction of secretory cell metaplasia and emphysema in hamsters, with *in vitro* observations on elastolysis by these enzymes. *Am Rev Respir Dis* 132: 362–366
10 Snider GL, Lucey EC, Christensen TG, Stone PJ, Calore JD, Catanese A, Franzblau C (1984) Emphysema and bronchial secretory cell metaplasia induced in hamsters by human neutrophil products. *Am Rev Respir Dis* 129: 155–160
11 Sommerhoff CP, Nadel JA, Basbaum CB, Caughey GH (1990) Neutrophil elastase and cathepsin G stimulate secretion from cultured bovine airway gland serous cells. *J Clin Invest* 85: 682–689
12 Amitani R, Wilson R, Rutman A, Read R, Ward C, Burnett D, Stockley RA, Cole PJ (1991) Effect of human neutrophil elastase and *Pseudomonas aeruginosa* proteinases on human respiratory epithelium. *Am J Respir Cell Mol Biol* 4: 26–32
13 Smallman LA, Hill SL, Stockley RA (1984) Reduction of ciliary beat frequency *in vitro* by sputum from patients with bronchiectasis: a serine proteinase effect. *Thorax* 39: 663–667
14 Stockley RA (2002) Neutrophils and the pathogenesis of COPD. *Chest* 121: 151S–155S
15 Hunninghake GW and Crystal RG (1983) Cigarette smoking and lung destruction. Accumulation of neutrophils in the lungs of cigarette smokers. *Am Rev Respir Dis* 128: 833–838
16 Betsuyaku T, Nishimura M, Takeyabu K, Tanino M, Venge P, Xu S, Kawakami Y (1999) Neutrophil granule proteins in bronchoalveolar lavage fluid from subjects with subclinical emphysema. *Am J Respir Crit Care Med* 159: 1985–1991
17 Damiano VV, Tsang A, Kucich U, Abrams WR, Rosenbloom J, Kimbel P, Fallahnejad M, Weinbaum G (1986) Immunolocalization of elastase in human emphysematous lungs. *J Clin Invest* 78: 482–493

18 Takeyama K, Agustí C, Ueki IF, Lausier J, Cardell LO, Nadel JA (1998) Neutrophil-dependent goblet cell degranulation: role of membrane-bound elastase and adhesion molecules. *Am J Physiol* 19: L294–L302
19 Takeyama K, Dabbagh K, Lee H-M, Agustí C, Lausier JA, Ueki IF, Grattan KM, Nadel JA (1999) Epidermal growth factor system regulates mucin production in airways. *Proc Natl Acad Sci USA* 96: 3081–3086
20 Finlay GA, Russell KJ, McMahon KJ, D'Arcy EM, Masterson JB, Fitzgerald MX, O'Connor CM (1997) Elevated levels of matrix metalloproteinases in bronchoalveolar lavage fluid of emphysematous patients. *Thorax* 52: 502–506
21 Hill AT, Campbell EJ, Hill SL, Bayley DL, Stockley RA (2000) Association between airway bacterial load and markers of airway inflammation in patients with stable chronic bronchitis. *Am J Med* 109: 288–295
22 Stockley RA and Burnett D (1979) Alpha$_1$-antitrypsin and leukocyte elastase in infected and non-infected sputum. *Am Rev Respir Dis* 120: 1081–1086
23 Dowson LJ, Guest PJ, Stockley RA (2001) Longitudinal changes in physiological, radiological, and health status measurements in alpha(1)-antitrypsin deficiency and factors associated with decline. *Am J Respir Crit Care Med* 164: 1805–1809
24 Donaldson GC, Seemungal TA, Bhowmik A, Wedzicha JA (2002) Relationship between exacerbation frequency and lung function decline in chronic obstructive pulmonary disease. *Thorax* 57: 847–852
25 Tanino M, Betsuyaku T, Takeyabu K, Tanino Y, Yamaguchi E, Miyamoto K, Nishimura M (2002) Increased levels of interleukin-8 in BAL fluid from smokers susceptible to pulmonary emphysema. *Thorax* 57: 405–411
26 McElvaney NG, Nakamura H, Birrer P, Hebert CA, Wong WL, Alphonso M, Baker JB, Catalano MA (1992) Modulation of airway inflammation in cystic fibrosis: *in vivo* suppression of interleukin-8 levels on the respiratory epithelial surface by aerosolization of recombinant secretory leukoprotease inhibitor. *J Clin Invest* 90: 1296–1301
27 Hubbard RC, Fells G, Gadek J, Pacholok S, Humes J, Crystal RG (1991) Neutrophil accumulation in the lung in α_1-antitrypsin deficiency: spontaneous release of leukotriene B4 by alveolar macrophages. *J Clin Invest* 88: 891–897
28 Crooks SW, Bayley DL, Hill SL, Stockley RA (2000) Bronchial inflammation in acute bacterial exacerbations of chronic bronchitis: the role of leukotriene B4. *Eur Respir J* 15: 274–280
29 Di Stefano AP, Maestrelli A, Roggeri G, Turato G, Calabro S, Potena A, Mapp CE, Ciaccia A, Covacev L, Fabbri LM et al (1994) Upregulation of adhesion molecules in the bronchial mucosa of subjects with chronic obstructive bronchitis. *Am J Respir Crit Care Med* 149: 803–810
30 Noguera A, Busquets X, Sauleda J, Villaverde JM, Macnee W, Agusti AG (1998) Expression of adhesion molecules and G proteins in circulating neutrophils in chronic obstructive pulmonary disease. *Am J Respir Crit Care Med* 158: 1664–1668
31 Burnett D, Chamba A, Hill SL, Stockley RA (1987) Neutrophils from subjects with

chronic obstructive lung disease show enhanced chemotaxis and extracellular proteolysis. *Lancet* 2: 1043–1046

32 Hill AT, Bayley DL, Campbell EJ, Hill SL, Stockley RA (2000) Airways inflammation in chronic bronchitis: the effects of smoking and alpha1-antitrypsin deficiency. *Eur Respir J* 15: 886–890

33 Crooks SW, Stockley RA (1998) Leukotriene B4. *Int J Biochem Cell Biol* 30: 173–178

34 Cassatella MA, Bazzoni F, Ceska M, Ferro I, Baggiolini M, Berton G (1992) IL-8 production by human polymorphonuclear leukocytes. The chemoattractant formyl-methionyl-leucyl-phenylalanine induces the gene expression and release of IL-8 through a pertussis toxin-sensitive pathway. *J Immunol* 148: 3216–3220

35 Liou TG and Campbell EJ (1996) Quantum proteolysis resulting from release of single granules by human neutrophils: a novel, nonoxidative mechanism of extracellular proteolytic activity. *J Immunol* 157: 2624–2631

36 Campbell EJ, Campbell MA, Boukedes SS, Owen CA (1999) Quantum proteolysis by neutrophils: implications for pulmonary emphysema in alpha(1)-antitrypsin deficiency. *J Clin Invest* 104: 337–344

37 Owen CA, Campbell MA, Sannes PL, Boukedes SS, Campbell EJ (1995) Cell surface-bound elastase and cathepsin G on human neutrophils: a novel, non-oxidative mechanism by which neutrophils focus and preserve catalytic activity of serine proteinases. *J Cell Biol* 131: 775–789

38 Lomas DA, Stone SR, Llewellyn-Jones C, Keogan MT, Wang ZM, Rubin H, Carrell RW, Stockley RA (1995) The control of neutrophil chemotaxis by inhibitors of cathepsin G and chymotrypsin. *J Biol Chem* 270: 23437–23443

39 Clark RA, Kimball HR (1971) Defective granulocyte chemotaxis in the Chediak-Higashi syndrome. *J Clin Invest* 50: 2645–2652

40 Shiflett SL, Kaplan J, Ward DM (2002) Chediak-Higashi syndrome: a rare disorder of lysosomes and lysosome-related organelles. *Pigment Cell Res* 15: 251–257

41 Kjeldsen L, Calafat J, Borregaard N (1998) Giant granules of neutrophils in Chediak-Higashi syndrome are derived from azurophil granules but not from specific and gelatinase granules. *J Leukoc Biol* 64: 72–77

42 Ward CJ (1994) The control of production of human leucocyte elastase (HLE) in mature neutrophils, cell lines and bone marrow cells. Thesis submitted for the degree of Doctor of Philosophy, University of Birmingham, UK

43 Tanaka E, Yuba Y, Sato A, Kuze F (1994) Effects of the beige mutation on respiratory tract infection with *Pseudomonas aeruginosa* in mice. *Exp Lung Res* 20: 351–366

44 Cavarra E, Martorana PA, Cortese S, Gambelli F, Di Simplicio P, Lungarella G (1997) Neutrophils in beige mice secrete normal amounts of cathepsin G and a 46 kDa latent form of elastase that can be activated extracellularly by proteolytic activity. *Biol Chem* 378: 417–423

45 Cavarra E, Martorana PA, de Santi M, Bartalesi B, Cortese S, Gambelli F, Lungarella G (1999) Neutrophil influx into the lungs of beige mice is followed by elastolytic damage and emphysema. *Am J Respir Cell Mol Biol* 20: 264–269

46 Toomes C, James J, Wood AJ, Wu CL, McCormick D, Lench N, Hewitt C, Moynihan L, Roberts E, Woods CG et al (1999) Loss-of-function mutations in the cathepsin C gene result in periodontal disease and palmoplantar keratosis. *Nature Genetics* 23: 421–424

47 Pham CTN, Ley TJ (1999) Dipeptidyl peptidase I is required for the processing and activation of granzymes A and B *in vivo*. *Proc Natl Acad Sci USA* 96: 8627–8632

48 Liu R, Cao C, Meng H, Tang Z (2000) Leukocyte functions in 2 cases of Papillon-Lefèvre syndrome. *J Clin Periodontol* 27: 69–73

49 Shapiro SD (2002) Neutrophil Elastase. Path clearer, pathogen killer, or just pathologic? *Am J Respir Cell Mol Biol* 26: 266–268

50 MacIvor DM, Shapiro SD, Pham CTN, Belaaouaj A, Abraham SN, Ley TJ (1999) Normal neutrophil function in cathepsin G-deficient mice. *Blood* 94: 4282–4293

51 Tkalcevic J, Novelli M, Phylactides M, Iredale JP, Segal AW, Roes J (2000) Impaired immunity and enhanced resistance to endotoxin in the absence of neutrophil elastase and cathepsin G. *Immunity* 12: 201–210

52 Belaaouaj A, McCarthy R, Baumann M, Gao Z, Ley TJ, Abraham SN, Shapiro SD (1998) Mice lacking neutrophil elastase reveal impaired host defence against gram negative bacterial sepsis. *Nature Medicine* 4: 615–618

53 Mackarel AJ, Cottell DC, Russell KJ, Fitzgerald MX, O'Connor CM (1999) Migration of neutrophils across human pulmonary endothelial cells is not blocked by matrix metalloproteinases or serine protease inhibitors. *Am J Respir Cell Mol Biol* 20: 1209–1219

54 Allport J, Ding H, Ager A, Steeber D, Tedder T, Luscinskas F (1997) L-selectin shedding does not regulate human neutrophil attachment, rolling or transmigration across human vascular endothelium *in vitro*. *J Immunol* 158: 4365–4372

55 Huber AR, Weiss SJ (1989) Disruption of the subendothelial basement membrane during neutrophil diapedesis in an *in vitro* construct of a blood vessel wall. *J Clin Invest* 83: 1122–1136

56 Delacourt C, Hérigault S, Delclaux C, Poncin A, Levame M, Harf A, Saudubray F, Lafuma C (2002) Protection against acute lung injury by intravenous or intratracheal pretreatment with EPI-HNE-4, a new potent neutrophil elastase inhibitor. *Am J Respir Cell Mol Biol* 26: 290–297

57 Stockley RA, Shaw J, Afford SC, Morrison HM, Burnett D (1990) Effect of alpha-1-proteinase inhibitor on neutrophil chemotaxis. *Am J Respir Cell Mol Biol* 2: 163–170

58 Witko-Sarsat V, Halbwachs-Mecarelli L, Schuster A, Nusbaum P, Ueki I, Canteloup S, Lenoir G, Descamps-Latscha B, Nadel JA (1999) Proteinase 3, a potent secretagogue in airways, is present in cystic fibrosis sputum. *Am J Respir Cell Mol Biol* 20: 729–736

59 Kao RC, Wehner NG, Skubitz KM, Gray BH, Hoidal JR (1988) Proteinase 3. A distinct human polymorphonuclear leukocyte proteinase that produces emphysema in hamsters. *J Clin Invest* 82: 1963–1973

60 Ludemann J, Utecht B, Gross WL (1990) Anti-neutrophil cytoplasm antibodies in Wegener's granulomatosis recognize an elastinolytic enzyme. *J Exp Med* 171: 357–362

61 Porges AJ, Redecha PB, Kimberly WT, Csernok E, Gross WL, Kimberly RP (1994) Anti-

neutrophil cytoplasmic antibodies engage and activate human neutrophils *via* Fc gamma RIIa. *J Immunol* 153: 1271–1280

62 Rooney CP, Taggart C, Coakley R, McElvaney NG, O'Neill SJ (2001) Anti-proteinase 3 antibody activation of neutrophils can be inhibited by alpha1-antitrypsin. *Am J Respir Cell Mol Biol* 24: 747–754

63 Elzouki AN, Segelmark M, Wieslander J, Eriksson S (1994) Strong link between the alpha 1-antitrypsin PiZ allele and Wegener's granulomatosis. *J Intern Med* 236: 543–548

64 Segelmark M, Elzouki AN, Wieslander J, Eriksson S (1995) The PiZ gene of alpha 1-antitrypsin as a determinant of outcome in PR3-ANCA-positive vasculitis. *Kidney Int* 48: 844–850

65 Miller HRP, Pemberton AD (2002) Tissue-specific expression of mast cell granule serine proteinases and their role in inflammation in the lung and gut. *Immunology* 105: 375–390

66 Lesser M, Padilla ML, Cardozo C (1992) Induction of emphysema in hamsters by intratracheal instillation of cathepsin B. *Am Rev Respir Dis* 145: 661–668

67 Cardozo C, Padilla ML, Choi HS, Lesser M (1992) Goblet cell hyperplasia in large intrapulmonary airways after intratracheal injection of cathepsin B into hamsters. *Am Rev Respir Dis* 145: 675–679

68 Burnett D, Crocker J, Stockley RA (1983) Cathepsin B-like cysteine proteinase activity in sputum and immunohistologic identification of cathepsin B in alveolar macrophages. *Am Rev Respir Dis* 128: 915–919

69 Burnett D, Abrahamson M, Devalia JL, Sapsford RJ, Davies RJ, Buttle DJ (1995) Synthesis and secretion of procathepsin B and cystatin C by human bronchial epithelial cells *in vitro*: modulation of cathepsin B activity by neutrophil elastase. *Arch Biochem Biophys* 317: 305–310

70 Buttle DJ, Abrahamson M, Burnett D, Mort JS, Barrett AJ, Dando PM, Hill SL (1991) Human sputum cathepsin B degrades proteoglycan, is inhibited by alpha 2-macroglobulin and is modulated by neutrophil elastase cleavage of cathepsin B precursor and cystatin C. *Biochem J* 276: 325–331

71 Taggart CC, Lowe GJ, Greene CM, Mulgrew AT, O'Neill SJ, Levine RL, McElvaney NG (2001) Cathepsin B, L, and S cleave and inactivate secretory leucoprotease inhibitor. *J Biol Chem* 276: 33345–33352

72 Johnson DA, Barrett AJ, Mason RW (1986) Cathepsin L inactivates alpha 1-proteinase inhibitor by cleavage in the reactive site region. *J Biol Chem* 261: 14748–14751

73 Mason RW, Johnson DA, Barrett AJ, Chapman HA (1986) Elastinolytic activity of human cathepsin L. *Biochem J* 233: 925–927

74 Shi GP, Munger JS, Meara JP, Rich DH, Chapman HA (1992) Molecular cloning and expression of human alveolar macrophage cathepsin S, an elastinolytic cysteine protease. *J Biol Chem* 267: 7258–7262

75 Takeyabu K, Betsuyaku T, Nishimura M, Yoshioka A, Tanino M, Miyamoto K,

Kawakami Y (1998) Cysteine proteinases and cystatin C in bronchoalveolar lavage fluid from subjects with subclinical emphysema. *Eur Respir J* 12: 1033–1039

76 Hautamaki RD, Kobayashi DK, Senior RM, Shapiro SD (1997) Requirement for macrophage elastase for cigarette smoke-induced emphysema in mice. *Science* 277: 2002–2004

77 Shapiro SD (2000) Animal models for COPD. *Chest* 117: 223S–227S

78 Churg A, Zay K, Shay S, Xie C, Shapiro SD, Hendricks R, Wright JL (2002) Acute cigarette smoke-induced connective tissue breakdown requires both neutrophils and macrophage metalloelastase in mice. *Am J Respir Cell Mol Biol* 27: 368–374

79 Wright JL, Farmer SG, Churg A (2002) Synthetic serine elastase inhibitor reduces cigarette smoke-induced emphysema in guinea pigs. *Am J Respir Crit Care Med* 166: 954–960

80 Dhami R, Gilks B, Xie C, Zay K, Wright JL, Churg A (2000) Acute cigarette smoke-induced connective tissue breakdown is mediated by neutrophils and prevented by α_1-antitrypsin. *Am J Respir Cell Mol Biol* 22: 244–252

81 Chandler S, Cossins J, Lury J, Wells G (1996) Macrophage metalloelastase degrades matrix and myelin proteins and processes a tumour necrosis factor-alpha fusion protein. *Biochem Biophys Res Commun* 228: 421–429

82 Liu Z, Zhou X, Shapiro SD, Shipley JM, Twining SS, Diaz LA, Senior RM, Werb Z (2000) The serpin α_1-proteinase inhibitor is a critical substrate for gelatinase B/MMP-9 *in vivo*. *Cell* 102: 647–655

83 Itoh Y and Nagase H (1995) Preferential inactivation of tissue inhibitor of metalloproteinases-1 that is bound to the precursor of matrix metalloproteinase 9 (progelatinase B) by human neutrophil elastase. *J Biol Chem* 270: 16518–16521

84 Ferry G, Lonchampt M, Pennel L, de Nanteuil G, Canet E, Tucker GC (1997) Activation of MMP-9 by neutrophil elastase in an *in vivo* model of acute lung injury. *FEBS Lett* 402: 111–115

85 Okada Y, Nakanishi I (1989) Activation of matrix metalloproteinase 3 (stromelysin) and matrix metalloproteinase 2 ('gelatinase') by human neutrophil elastase and cathepsin G. *FEBS Lett* 249: 353–356

86 Churg A, Wang RD, Tai H, Wang X, Xie C, Dai J, Shapiro SD, Wright JL (2003) Macrophage metalloelastase mediates acute cigarette smoke-induced inflammation *via* TNFα release. *Am J Respir Crit Care Med* 167: 1083–1089

87 D'Armiento J, Dalal S, Okada Y, Berg R, Chada K (1992) Collagenase expression in the lung of transgenic mice causes pulmonary emphysema. *Cell* 71: 955–961

88 Finlay GA, O'Driscoll LR, Russell KJ, D'Arcy EM, Masterson JB, Fitzgerald MX, O'Connor CM (1997) Matrix metalloproteinase expression and production by alveolar macrophages in emphysema. *Am J Respir Crit Care Med* 156: 240–247

89 Segura-Valdez L, Pardo A, Gaxiola M, Uhal BD, Becerril C, Selman M (2000) Upregulation of gelatinases A and B, collagenases 1 and 2, and increased parenchymal cell death in COPD. *Chest* 117: 684–694

90 Ohnishi K, Takagi M, Kurokawa Y, Satomi S, Konttinen YT (1998) Matrix metallo-

proteinase-mediated extracellular matrix protein degradation in human pulmonary emphysema. *Lab Invest* 78: 1077–1087
91 Imai K, Dalal SS, Chen ES, Downey R, Schulman LL, Ginsburg D, D'Armiento J (2001) Human collagenase (matrix metalloproteinase-1) expression in the lungs of patients with emphysema. *Am J Respir Crit Care Med* 163: 786–791
92 Vignola AM, Bonanno A, Mirabella A, Riccobono L, Mirabella F, Profita M, Bellia V, Bousquet J, Bonsignore G (1998) Increased levels of elastase and alpha1-antitrypsin in sputum of asthmatic patients. *Am J Respir Crit Care Med* 157: 505–511
93 Cataldo D, Munaut C, Noël A, Frankenne F, Bartsch P, Foidart JM, Louis R (2001) Matrix metalloproteinases and TIMP-1 production by peripheral blood granulocytes from COPD patients and asthmatics. *Allergy* 56: 145–151
94 Lim S, Roche N, Oliver BG, Mattos W, Barnes PJ, Chung KF (2000) Balance of matrix metalloprotease-9 and tissue inhibitor of metalloprotease-1 from alveolar macrophages in cigarette smokers. Regulation by interleukin-10. *Am J Respir Crit Care Med* 162: 1355–1360
95 Russell RE, Culpitt SV, DeMatos C, Donnelly L, Smith M, Wiggins J, Barnes PJ (2002) Release and activity of matrix metalloproteinase-9 and tissue inhibitor of metalloproteinase-1 by alveolar macrophages from patients with chronic obstructive pulmonary disease. *Am J Respir Cell Mol Biol* 26: 602–609
96 Minematsu N, Nakamura H, Tateno H, Nakajima T, Yamaguchi K (2001) Genetic polymorphism in matrix metalloproteinase-9 and pulmonary emphysema. *Biochem Biophys Res Commun* 289: 116–119
97 Joos L, He JQ, Shepherdson MB, Connett JE, Anthonisen NR, Pare PD, Sandford AJ (2002) The role of matrix metalloproteinase polymorphisms in the rate of decline in lung function. *Hum Mol Genet* 11: 569–576
98 Hirano K, Sakamoto T, Uchida Y, Morishima Y, Masuyama K, Ishii Y, Nomura A, Ohtsuka M, Sekizawa K (2001) Tissue inhibitor of metalloproteinases-2 gene polymorphisms in chronic obstructive pulmonary disease. *Eur Respir J* 18: 748–752
99 Wan H, Winton HL, Soeller C, Tovey ER, Gruenert DC, Thompson PJ, Stewart GA, Taylor GW, Garrod DR, Cannell MB, Robinson C (1999) Der p 1 facilitates transepithelial allergen delivery by disruption of tight junctions. *J Clin Invest* 104: 123–133
100 Kalsheker NA, Deam S, Chambers L, Sreedharan S, Brocklehurst K, Lomas DA (1996) The house dust mite allergen Der p1 catalytically inactivates α_1-antitrypsin by specific reactive centre loop cleavage: A mechanism that promotes airway inflammation and asthma. *Biochem Biophys Res Comm* 221: 59–61
101 Hewitt CR, Brown AP, Hart BJ, Pritchard DI (1995) A major house dust mite allergen disrupts the immunoglobulin E network by selectively cleaving CD23: innate protection by antiproteases. *J Exp Med* 182: 1537–1544
102 Hendrixson DR, St Geme JWI (1998) The *Haemophilus influenzae* Hap serine protease promotes adherence and microcolony formation, potentiated by a soluble host protein. *Mol Cell* 2: 841–850
103 Bieth JG (1986) Elastases: catalytic and biological properties. In: RP Mecham (ed): *Biol-*

ogy of extracellular matrix: regulation of matrix accumulation. Academic Press, Orlando, 217–320
104 Niederman MS, Merrill WW, Polomski LM, Reynolds HY, Gee JB (1986) Influence of sputum IgA and elastase on tracheal cell bacterial adherence. *Am Rev Respir Dis* 133: 255–260
105 Doring G, Frank F, Boudier C, Herbert S, Fleischer B, Bellon G (1995) Cleavage of lymphocyte surface antigens CD2, CD4, and CD8 by polymorphonuclear leukocyte elastase and cathepsin G in patients with cystic fibrosis. *J Immunol* 154: 4842–4850
106 Bank U, Kupper B, Reinhold D, Hoffmann T, Ansorge S (1999) Evidence for a crucial role of neutrophil-derived serine proteases in the inactivation of interleukin-6 at sites of inflammation. *FEBS Lett* 461: 235–240
107 Okada Y, Watanabe S, Nakanishi I, Kishi J, Hayakawa T, Watorek W, Travis J, Nagase H (1988) Inactivation of tissue inhibitor of metalloproteinases by neutrophil elastase and other serine proteinases. *FEBS Lett* 229: 157–160
108 Taipale J, Lohi J, Saarinen J, Kovanen PT, Keski-Oja J (1995) Human mast cell chymase and leukocyte elastase release latent transforming growth factor-β1 from the extracellular matrix of cultured human epithelial and endothelial cells. *J Biol Chem* 270: 4689–4696
109 Rice A, Banda MJ (1995) Neutrophil elastase processing of gelatinase A is mediated by extracellular matrix. *Biochemistry* 34: 9249–9256
110 Fick RBJ, Robbins RA, Squier SU, Schoderbek WE, Russ WD (1986) Complement activation in cystic fibrosis respiratory fluids: *in vivo* and *in vitro* generation of C5a and chemotactic activity. *Pediatr Res* 20: 1258–1268
111 Vogt W (2000) Cleavage of the fifth component of complement and generation of a functionally active C5b6-like complex by human leukocyte elastase. *Immunobiology* 201: 470–477
112 Joslin G, Griffin GL, August AM, Adams S, Fallon RJ, Senior RM, Perlmutter DH (1992) The serpin-enzyme complex (SEC) receptor mediates the neutrophil chemotactic effect of alpha-1 antitrypsin-elastase complexes and amyloid-beta peptide. *J Clin Invest* 90: 1150–1154
113 Perlmutter DH, Travis J, Punsal PI (1988) Elastase regulates the synthesis of its inhibitor, alpha 1-proteinase inhibitor, and exaggerates the defect in homozygous PiZZ alpha 1 PI deficiency. *J Clin Invest* 81: 1774–1780
114 Perlmutter DH, Glover GI, Rivetna M, Schasteen CS, Fallon RJ (1990) Identification of a serpin-enzyme complex receptor on human hepatoma cells and human monocytes. *Proc Natl Acad Sci USA* 87: 3753–3757
115 Nahori MA, Renesto P, Vargaftig BB, Chignard M (1992) Activation and damage of cultured airway epithelial cells by human elastase and cathepsin G. *Eur J Pharmacol* 228: 213–218
116 Walsh DE, Greene CM, Carroll TP, Taggart CC, Gallagher PM, O'Neill SJ, McElvaney NG (2001) Interleukin-8 up-regulation by neutrophil elastase is mediated by MyD88/IRAK/TRAF-6 in human bronchial epithelium. *J Biol Chem* 276: 35494–35499

117 Sallenave JM, Shulmann J, Crossley J, Jordana M, Gauldie J (1994) Regulation of secretory leukocyte proteinase inhibitor (SLPI) and elastase-specific inhibitor (ESI/elafin) in human airway epithelial cells by cytokines and neutrophilic enzymes. *Am J Respir Cell Mol Biol* 11: 733–741

118 van Wetering S, van der Linden AC, van Sterkenburg MA, Rabe KF, Schalkwijk J, Hiemstra PS (2000) Regulation of secretory leukocyte proteinase inhibitor (SLPI) production by human bronchial epithelial cells: increase of cell-associated SLPI by neutrophil elastase. *J Invest Med* 48: 359–366

119 Reid PT, Marsden ME, Cunningham GA, Haslett C, Sallenave JM (1999) Human neutrophil elastase regulates the expression and secretion of elafin (elastase-specific inhibitor) in type II alveolar epithelial cells. *FEBS Lett* 457: 33–37

120 Voynow JA, Young LR, Wang Y, Horger T, Rose MC, Fischer BM (1999) Neutrophil elastase increases MUC5AC mRNA and protein expression in respiratory epithelial cells. *Am J Physiol* 276: L835–L843

121 Smedly LA, Tonnesen MG, Sandhaus RA, Haslett C, Guthrie LA, Johnston RB Jr, Henson PM, Worthen GS (1986) Neutrophil-mediated injury to endothelial cells. Enhancement by endotoxin and essential role of neutrophil elastase. *J Clin Invest* 77: 1233–1243

122 Belaaouaj A (2002) Neutrophil elastase-mediated killing of bacteria: lessons from targeted mutagenesis. *Microbes Infect* 4: 1259–1264

123 Dal Nogare AR, Toews GB, Pierce AK (1987) Increased salivary elastase precedes gram-negative bacillary colonization in postoperative patients. *Am Rev Respir Dis* 135: 671–675

124 Plotkowski MC, Beck G, Tournier JM, Bernardo-Filho M, Marques EA, Puchelle E (1989) Adherence of *Pseudomonas aeruginosa* to respiratory epithelium and the effect of leucocyte elastase. *J Med Microbiol* 30: 285–293

125 Woods DE (1987) Role of fibronectin in the pathogenesis of gram-negative bacillary pneumonia. *Rev Infect Dis* 9 (Suppl 4): S386–S390

126 Abbinante-Nissen JM, Simpson LG, Leikauf GD (1993) Neutrophil elastase increases secretory leukocyte protease inhibitor transcript levels in airway epithelial cells. *Am J Physiol* 265: L286–L292

127 Nakamura H, Yoshimura K, McElvaney NG, Crystal RG (1992) Neutrophil elastase in respiratory epithelial lining fluid of individuals with cystic fibrosis induces interleukin-8 gene expression in a human bronchial epithelial cell line. *J Clin Invest* 89: 1478–1484

128 Sternlicht MD, Werb Z (2001) How matrix metalloproteinases regulate cell behaviour. *Annu Rev Cell Dev Biol* 17: 463–516

Mucus hypersecretion in COPD

Duncan F. Rogers

Thoracic Medicine, National Heart & Lung Institute, Imperial College, Dovehouse Street, London SW3 6LY, UK

Introduction

Airway mucus hypersecretion is a cardinal feature of chronic obstructive pulmonary disease (COPD) (Fig. 1). Or is it? In fact, mucus hypersecretion, implicit in the term chronic bronchitis, is one of three pathophysiological entities comprising COPD, the other two being chronic bronchiolitis (small airways disease) and emphysema (alveolar destruction). The relative contribution of each component to pathophysiology varies between patients, with the impact of mucus hypersecretion on morbidity and mortality varying accordingly. Thus, although previously included in definitions of COPD [1], the term "mucus hypersecretion" is omitted from current definitions [2]. Nevertheless, there are many patients in whom hypersecretion has clinical significance, for example patients prone to chest infections [3]. Consequently, it is important to understand the mechanisms underlying mucus hypersecretion in COPD. This in turn should allow identification of pathophysiological targets and rational development of pharmacotherapeutic drugs. The present chapter: 1) assesses the role of airway mucus hypersecretion in the pathophysiology of the "bronchitic" component of COPD, 2) considers the clinical impact of mucus hypersecretion in COPD, and 3) discusses potential novel therapy for this condition. The chapter begins with a brief overview of airway mucus and mucins.

Airway mucus and mucins

In health, a thin film of slimy liquid protects the airway surface [4]. The liquid is often referred to as 'mucus' and is a complex 1–2% aqueous solution of salts, enzymes and anti-enzymes, oxidants and antioxidants, bacterial products, antibacterial agents, cell-derived mediators and proteins, plasma-derived mediators and proteins, and cell debris such as DNA. The mucus forms an upper gel layer and a lower sol layer. Inhaled particles are trapped in the gel layer and, by transportation on the tips of beating cilia, are removed from the airways, a process termed mucocil-

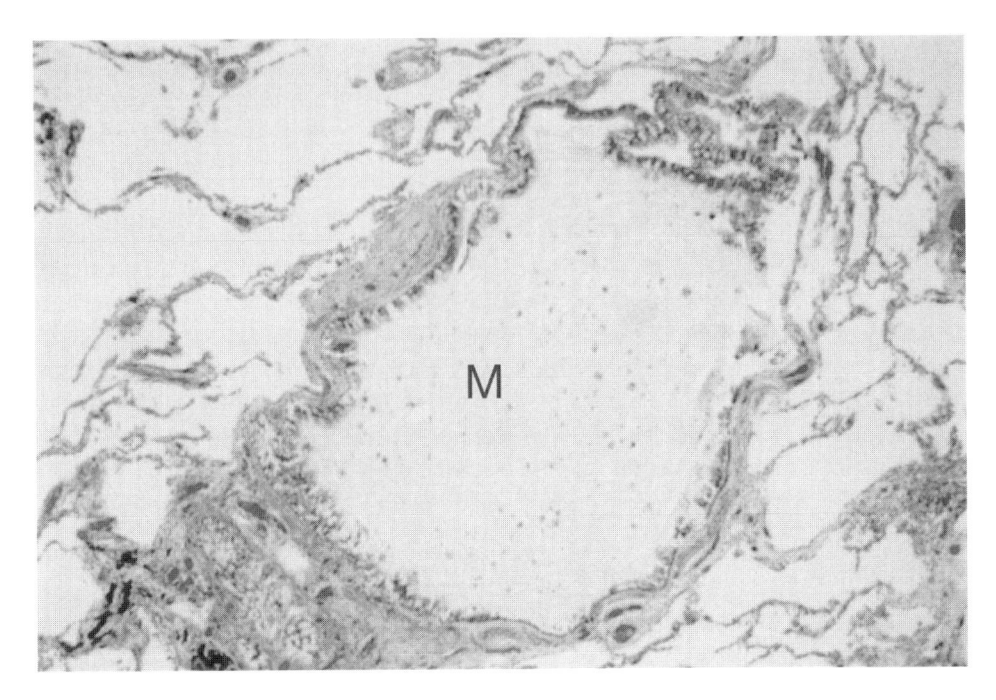

Figure 1
Mucus hypersecretion in COPD. Mucus (M) occluding a small airway in a cigarette smoker with chronic sputum production.

iary clearance. Airway mucus requires an optimal combination of viscosity and elasticity for efficient ciliary interaction. Viscoelasticity is conferred primarily by high molecular weight mucins that comprise up to 2% by weight of the mucus [5]. Airway mucins are produced by epithelial goblet cells and submucosal glands. Mature mucins are long thread-like molecules composed of monomers joined end-to-end by disulphide bridges. The monomers comprise a highly glycosylated linear peptide sequence, termed apomucin, that is encoded by specific mucin (MUC) genes. Seventeen human MUC genes are reported to date, namely MUC1, 2, 3A, 3B, 4, 5AC, 5B, 6–9, 11–13 and 16–18 [5–9]. The MUC5AC and MUC5B gene products are the major gel forming mucins in "normal" respiratory tract secretions [5].

Pathophysiology of mucus hypersecretion in COPD

The pathophysiology of mucus hypersecretion in COPD has characteristic features. Some features, such as sputum production, are shared with other hypersecretory respiratory diseases, for example asthma. Other features appear to be specific for

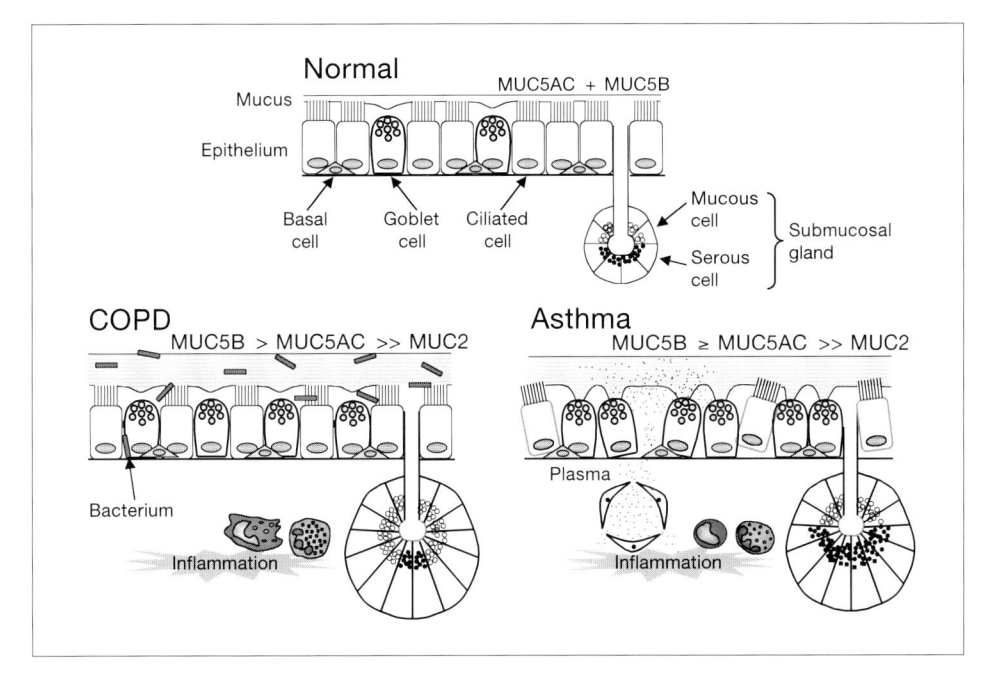

Figure 2
Differences in airway mucus pathophysiology between COPD and asthma. Compared with normal, in COPD there is increased luminal mucus, goblet cell hyperplasia, submucosal gland hypertrophy (with an increased proportion of mucous to serous acini), an increased ratio of mucin (MUC) 5B (low charge glycoform) to MUC5AC, small amounts of MUC2, and respiratory infection. Pulmonary inflammation includes macrophages and neutrophils. In asthma, there is increased luminal mucus, epithelial 'fragility', marked goblet cell hyperplasia, submucosal gland hypertrophy (although without an increased mucous to serous ratio), 'tethering' of mucus to goblet cells, and plasma exudation. Airway inflammation includes T lymphocytes and eosinophils. Many of these differences require more data from greater numbers of subjects.

COPD (see below). Differences in mucus pathophysiology between COPD and asthma have been discussed previously [10], and are summarised in Figure 2. In addition, the pulmonary inflammation of COPD (essentially a macrophage-driven neutrophilia) that induces the hypersecretory phenotype of COPD is different to asthma.

Sputum production, up to 100 ml per day in some patients, is associated with excessive airway mucus (Fig. 1) [11–13]. The increased mucus is associated with goblet cell hyperplasia [11, 14] and submucosal gland hypertrophy [11, 12, 15, 16]. In particular, gland mucous cells are markedly increased relative to serous cells [17]. This is in contrast to asthma where the glands, albeit hypertrophied, were otherwise morphologically normal. Gland size correlates with amount of luminal mucus and daily sputum volume [15]. Although not necessarily causal, the latter observation suggests a strong relationship between gland hypertrophy and mucus hypersecretion in COPD.

Not all COPD patients exhibit all of the above features of hypersecretion. Not every patient expectorates, and there is overlap in gland size with healthy non-smokers, and also between sputum producers and non-producers [14, 16, 18-20]. Although goblet cell hyperplasia is associated with degree of airway inflammation, gland size is not [14]. Goblet cell hyperplasia is not noted in all patients [12, 17]. Thus, although considered a general feature of COPD, mucus hypersecretion is not diagnostic in all cases.

The mucin composition of airway mucus in patients with COPD may be abnormal. Mucins in sputum are less acidic than normal [21], which may relate to altered glycosylation. MUC5AC and a low charge glycoform of MUC5B are the major mucin species in patients with COPD [22–25], with the low charge glycoform increased above normal levels [26]. The significance to bacterial colonisation of the change in MUC5B glycoforms in unclear. However, COPD patients are prone to respiratory infection [2].

In contrast to normal airways, goblet cells in COPD contain not only MUC5AC but also MUC5B [24, 27] and MUC2 [5]. This distribution is different to that in patients with asthma or cystic fibrosis (CF), where MUC5AC and MUC5B show a similar histological distribution to normal controls [28, 29]. Although not found consistently [13, 23], MUC2 may be increased in "irritated" airways, including COPD [5, 26, 30].

Cilia and mucociliary clearance in COPD

In addition to abnormalities in airway mucus in COPD, there are also abnormalities in ciliated cells and cilia. The number of ciliated cells and ciliary length is decreased in patients with chronic bronchitis [31]. Ciliary aberrations include compound cilia, cilia enclosed within periciliary sheaths, and cilia with abnormal axonemes or intra-cytoplasmic microtubule doublets [32]. These abnormalities coupled with mucus hypersecretion are associated with reduced mucus clearance and airway obstruction.

Airway mucus clearance is impaired in patients with COPD [33]. There are often discrepancies between studies, due invariably to differences in methodology [34], but may also be due to observations made at different stages of disease. The validity of studies in COPD is dependent upon patient selection and the exclusion of patients with asthma [35]. Lung clearance is significantly reduced in heavy smokers [36] and in patients with chronic bronchitis [37]. Forced expirations and cough compensate relatively effectively for decreased mucus clearance in patients with chronic bronchitis but not in patients with emphysema where lung elastic recoil is impaired [38, 39].

Epidemiology of mucus hypersecretion in COPD

The role of mucus in pathophysiology and clinical symptoms in COPD is controversial [10]. Epidemiological studies sampling hundreds to thousands of subjects in the late 1970s and '80s found scant evidence for the involvement of mucus in either the mortality or accelerated age-related decline in lung function associated with COPD [40–44]. In all studies, sputum production was the index of mucus hypersecretion. However, the relationship between sputum production and mucus hypersecretion, particularly in the small airways, the main site of airflow obstruction, is unclear. Nevertheless, the consensus of these studies was that airflow obstruction and mucus hypersecretion were largely independent disease processes.

In contrast, a number of studies in the late 1980s and '90s found positive associations between sputum production and decline in lung function, hospitalisation and death [45–49]. Some of these reports were re-examinations of the same patients, now older, reported previously. Of note is the observation that incidence of death was related to increased risk of patients with phlegm production to die of respiratory infection [3]. Thus, although not strongly associated with disease progression in all cases, mucus hypersecretion contributes to morbidity and mortality in certain groups of patients with COPD, particularly those prone to infection, and possibly as patients age. This highlights the importance of developing drugs that inhibit mucus hypersecretion in these patients.

Pharmacotherapy of mucus hypersecretion in COPD

The clinical symptoms of cough and sputum production, coupled with a perception of the importance of mucus hypersecretion in the pathophysiology of a number of severe lung conditions, including COPD, has prompted renewed interest in research into airway hypersecretion and, in concert, in development of drugs targeting mucus. It should be noted that COPD has specific trigger factors, "profile" of pulmonary inflammation and mucus hypersecretory phenotype (Fig. 2), and specific drugs may be required to fulfil the theoretical requirements for treatment of hypersecretion in COPD (Tab. 1). The following sections consider different approaches to inhibition of mucus hypersecretion in COPD, starting with suppression of lung inflammation, possibly the most beneficial therapy overall (Fig. 3). Other approaches include neural inhibition, inhibition of mucin secretion, and inhibition of MUC gene expression, mucin synthesis and goblet cell hyperplasia (Fig. 4). Finally, compounds altering mucus properties are discussed.

Table 1 - Theoretical requirements for pharmacotherapy of mucus pathophysiology in COPD

Overall effect	Component effects
Facilitate mucus clearance (short-term relief of symptoms)	Reduce viscosity (increase elasticity)
	Increase ciliary function
	Induce cough
Reverse hypersecretory phenotype (long-term benefit)	Reduce submucosal gland size
	Correct increased gland mucous:serous cell ratio
	Reduce goblet cell number
	Reverse increased lcgf-MUC5B:MUC5AC ratio

lcgf, low charge glycoform; MUC mucin gene product.

Anti-inflammatory therapy

Anti-inflammatory drugs should have beneficial effects on mucus hypersecretion in COPD. For example, glucocorticosteroids inhibit mucus secretion, goblet cell hyperplasia and MUC gene expression in experimental systems [50]. However, in contrast to asthma where they are clinically effective, in part due to an anti-hypersecretory action, glucocorticoids have limited effectiveness in stable COPD. In contrast, inhalation of the non-steroidal anti-inflammatory drug indomethacin, a cyclo-oxygenase (COX) inhibitor with no therapeutic benefit in asthma, reduces mucus output in patients with chronic bronchitis [51]. Selective COX-2 inhibitors with reduced gastrointestinal activity are on the market, including rofecoxib, celecoxib and eterocoxib, but not for COPD. The difference in efficacy of conventional anti-inflammatory drugs between COPD and asthma highlights the need to define the differences in inflammatory profile and hypersecretory phenotype between the two conditions.

Epoxygenase inducers

In addition to COX, cytochrome P-450 enzymes, termed epoxygenases, also metabolise arachidonic acid and regulate inflammation [52]. Epoxygenase activity catalyses the production of epoxy acids, hydroxy acids and hepoxilins. An inducer of epoxygenase, benzafibrate, inhibits airway goblet cell hyperplasia in a rat model of chronic bronchitis [53]. The mechanisms underlying the inhibition are unclear, but include production of anti-inflammatory mediators and reduction in amount of "available" arachidonic acid. Development of epoxygenase inducers, or of selective

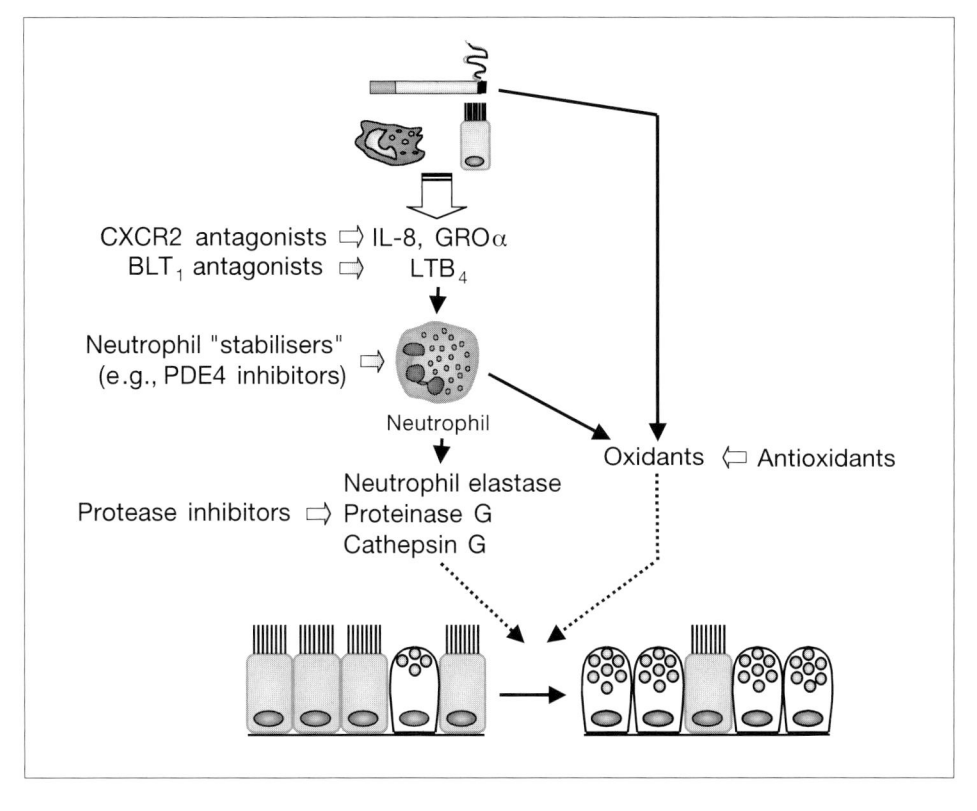

Figure 3
Relationship between neutrophilic inflammation in COPD and generation of a hypersecretory phenotype (e.g., goblet cell hyperplasia). Interactions between inhaled pollutants (e.g., cigarette smoke), macrophages and epithelial cells generate neutrophil chemoattractants, with resultant release of factors that induce mucus hypersecretion. The sequence of initiating events can be inhibited at different levels. PDE, phosphodiesterase.

eopxyeicosanoids, is a completely novel potential therapy for the inflammation and mucus hypersecretion of COPD.

Neutrophil inhibitors

In contrast to asthma, where most of the inflammatory mediators implicated in pathophysiology have effects on mucus [54], very few mediators in COPD have direct effects on mucus. However, neutrophil activation, with production of proteases and oxidants, induces mucin synthesis and goblet cell hyperplasia [55]. Consequently,

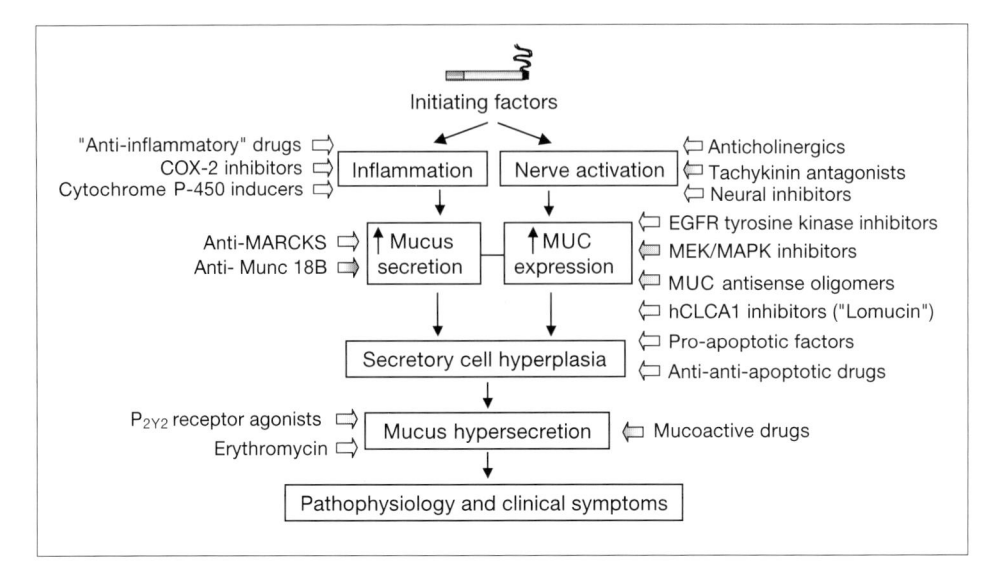

Figure 4
Pathophysiological "cascade" underlying mucus hypersecretion in COPD and sites of action of 'anti-hypersecretory' drugs (some compounds may act at more than one site, and the precise site of action of some compounds is unclear). CLCA, calcium activated chloride channel; COX, cyclo-oxygenase; EGFR, epidermal growth factor receptor; MAPK, mitogen activated protein kinase; MARCKS, myristoylated alanine-rich C kinase substrate; MEK, mitogen-activated protein kinase kinase; MUC, mucin (gene/protein).

inhibition of neutrophil influx to chemotactic factors such as LTB_4, IL-8 and GROα might have indirect beneficial effects on mucus hypersecretion in COPD. LTB_4 signals via BLT_1 receptors, and IL-8 and GROα (CXC chemokines) via the CXCR2 receptor. Small molecule antagonists at BLT_1 receptors, for example LY29311 and SB201146, and at CXCR2 receptors, for example SB265610, are in clinical development. Phosphodiesterase (PDE)-4 inhibitors also inhibit airway neutrophilia [56].

Protease inhibitors

Neutrophil elastase, cathepsin G and proteinase-3 have potent effects on airway mucus [57, 58]. Protease inhibitors may, therefore, inhibit mucus hypersecretion in COPD. Trifluoromethyl ketone human neutrophil elastase inhibitors such as ICI 200,355, inhibit neutrophil and elastase-mediated hypersecretory responses [59]. Other small molecule inhibitors such as ONO-5046 [60] are in development. Interestingly, the macrolide antibiotics erythromycin and flurythromycin have anti-neutrophil elastase activity, with erythromycin acting as a false substrate, and flurythromycin being an inactivator [61].

Inhibitors of reactive gas species

Oxidant stress is considered a pathophysiological feature of COPD [62]. In addition, exhaled nitric oxide (NO) is elevated in COPD [63]. Both oxidants and NO have significant effects on airway mucus and goblet cells [55,64]. Consequently, antioxidants and inhibitors of inducible NO synthase (iNOS) may have therapeutic benefit for mucus hypersecretion in COPD. N-acetylcysteine is a mucolytic agent that also has antioxidant properties, and is useful in exacerbations of COPD [65]. A variety of other classes of antioxidant are being investigated, including superoxide dismutase mimetics such as M40403 and spin trap compounds such as U101033E. Similarly, a variety of selective small molecule iNOS inhibitors are in development, including GW273629 and L-NIL.

Neural inhibitors

Neural mechanisms may contribute to the pathophysiology of mucus hypersecretion in COPD [66]. Neural influences can be suppressed by neurotransmitter receptor antagonists, inhibition of nerve activation and inhibition of neurotransmitter release.

Anticholinergics
Inhaled anticholinergics are conventional therapy in COPD [2]. Although administered as bronchodilators, part of their activity may be anti-secretory. Cholinergic nerve activity is the dominant drive to airway secretion, an effect mediated *via* muscarinic M_3 receptors on the secretory cells [66]. The autoinhibitory M_2 receptor regulates the magnitude of secretion. Non-selective muscarinic antagonists such as ipratropium bromide and oxitropium bromide have beneficial effects on airway mucus with concomitant, albeit modest, improvements in lung function [67, 68]. Anticholinergics with selectivity for the M_3 receptor over the M_2 receptor may have therapeutic benefit over non-selective compounds, for example tiotropium [69] and J104129 [70].

Tachykinin receptor antagonists
Neuronal C-fibres containing substance P (SP) and neurokinin A (NKA) form a sensory neural system with a motor function [66]. SP and NKA increase airway mucin secretion *via* interaction with tachykinin NK_1 receptors. Thus, activation of these

nerves might contribute to the pathophysiology of hypersecretion in COPD. Many tachykinin receptor antagonists are in development [71]. These include peptide and non-peptide antagonists selective for the NK_1, NK_2 or NK_3 receptor, as well as dual antagonists at NK_1 and NK_2 receptors. Few of these are in clinical trial for airway diseases, and the results from a limited number of trials in asthma have for the most part been inconclusive [66].

Inhibition of neurotransmitter release
Inhibition of both cholinergic and sensory nerves might be required for effective control of neurogenic hypersecretion, but would entail use of more than one antagonist. This could be resolved using a single inhibitor of neurotransmitter release, or of nerve activation. Neurotransmitter release from sensory and cholinergic nerves, with concomitant inhibition of neurogenic secretion, can be inhibited by activation of several types of prejunctional receptor, in particular μ and δ opioid receptors [66]. Cannabinoids also inhibit acetylcholine release [72] *via* interaction with prejunctional CB_2 receptors (rather than the CB_1 receptor that mediates the central effects of cannabinoids). CB_2 selective agonists, such as AM1241 and SR144528, might, therefore, be useful in reducing neurogenic airway hypersecretion.

Many of the prejunctional inhibitory receptors on airway nerves work *via* a common mechanism that involves opening of large conductance calcium-activated potassium (BK_{Ca}) channels [66]. Drugs that open BK_{Ca} channels such as NS1619 are, therefore, potential treatments for neurogenic hypersecretion, but to date no clinical studies are reported.

Inhibition of nerve activation
The vanilloid VR-1 receptor mediates activation of sensory nerves to a variety of stimulants [73]. Selective VR-1 antagonists, such as capsazepine, are in development. Cannabinoids also inhibit sensory nerve activation, including the endogenous cannabinoid anandamide, which also inhibits VR-1 [74].

Inhibition of secretion

Inhibition of secretion is a therapeutic option for mucus hypersecretion in COPD. However, inhibition of secretion could lead to excessive accumulation of intracellular mucins, with unknown, and potentially detrimental, effects on goblet cell function.

Myristoylated alanine-rich C kinase substrate (MARCKS)
MARCKS protein is a key intracellular molecule involved in intracellular movement

and exocytosis of mucin granules [75]. Blockade of MARCKS by a synthetic peptide to its N-terminal region inhibited mucin secretion by normal human bronchial epithelial cells *in vitro*. If MARCKS was a key secretory signalling molecule in humans *in vivo*, inhibition of MARCKS may be a therapeutic option for mucus hypersecretion in COPD.

Munc inhibitors

The Sec1/Munc18 family are critical to exocytosis in a variety of secretory cells including airway goblet cells. Intriguingly, experimental induction of Munc18B induces a marked airway hypersecretory phenotype [76]. Inhibition of Munc18B using antisense technology is being investigated with a view to inhibition of mucus hypersecretion.

Inhibition of goblet cell hyperplasia

Increased MUC gene expression, mucin synthesis and goblet cell hyperplasia appear to be linked processes that are regulated by a number of inflammatory mechanisms.

Epidermal growth factor receptor (EGF-R) tyrosine kinase inhibitors

Airway EGF-R expression is induced by a variety of experimental procedures pertinent to the pathophysiology of COPD including TNF-α, agarose plugs (mimicking airway irritation) and oxidative stress. EGF-R up-regulation, with signalling *via* EGF-R tyrosine kinase, appears to be a central signalling event for induction of mucin synthesis and goblet cell hyperplasia [55]. Inhibitors of EGF-R tyrosine kinase, for example AG1478, BIBX1522 and ZD1839 (Iressa), block these responses. Iressa is in clinical trial for cancer, but not yet for COPD or similar respiratory diseases.

Mitogen activated protein kinase (MAPK) inhibitors

The p38 MAP kinase and ERK MAP kinase pathways appear pivotal in a number of intracellular signalling pathways leading to mucin synthesis and goblet cell hyperplasia [77]. For example, the hypersecretory response to oxidative stress is blocked by a selective MAPK kinase (MEK) inhibitor, PD98059 [78]. This same inhibitor blocked MUC5AC expression induced *in vitro* by vanadium, an active component of the industrial air pollutant residual oil fly ash [79]. Small molecule inhibitors are in development, for example SB 203580 and SB 239063.

Calcium-activated chloride channel inhibitors

Calcium-activated chloride (CLCA) channels (originally reported as Gob-5) are another cellular moiety that appears to be critically involved in development of an airway hypersecretory phenotype. In mice, suppression of mCLCA3 inhibits goblet cell hyperplasia, whereas overexpression increases goblet cell number [80]. Lomucin (MSI 1956) is a small molecule putative inhibitor of hCLCA1 that is currently in clinical trial for hypersecretory airway diseases. The results of these trials are awaited with great interest.

Induction of goblet cell apoptosis
Hyperplastic airway goblet cells in COPD models express the antiapoptotic factor Bcl-2 [81]. Conversely, the pro-apoptotic factor Bax is crucial for resolution of hyperplasia [82]. Thus, the balance between Bcl-2 and Bax may determine maintenance of goblet cell hyperplasia. Targeted reduction of Bcl-2 expression by antisense oligonucleotides (ODN64 or ODN83] causes a dose-dependent resolution of hyperplasia.

Antisense oligonucleotides
Antisense technology is developing rapidly and oligomers for a number of gene products are already available. An antisense oligonucleotide to MARCKS downregulated both mRNA and protein levels and also attenuated mucin secretion [75]. In addition, an 18-mer MUC antisense oligomer suppressed mucin gene expression and wood smoke-induced epithelial metaplasia in rabbit airways [83].

Alteration of mucus properties

There are a number of approaches to alter the biophysical properties of mucus, in general to aid clearance.

Mucoactive drugs
Numerous compounds with potentially beneficial actions on mucus are listed in pharmacopoeias worldwide, including N-acetylcysteine, ambroxol, bromhexine and carbocysteine [65]. However, in contrast to their abundance and availability, mucolytic-mucoactive drugs are not generally recommended in management of COPD [2]. Inconsistency and imprecision in design of clinical trials has hindered the clinical evaluation of these compounds [84]. In phase I trial in patients with CF, Nacystelyn, a lysine derivative of N-acetylcysteine with greater mucolytic and antioxidant activity [85], has beneficial effects on airway mucus [86].

Purinoceptor P_{2Y2} receptor agonists and antagonists

The purine nucleotides, adenosine 5'-triphosphate (ATP) and uridine triphosphate (UTP), are stimulants of airway mucin and water secretion [87,88]. These activities are perceived to have therapeutic significance in hypersecretory disorders of the airways, including COPD, and are mediated *via* P_{2Y2} receptors. Consequently, P_{2Y2} antagonists might be effective in inhibiting airway hypersecretion. However, mucus hydration is associated with improvements in mucociliary clearance. Consequently, stimulation of water secretion is currently perceived to be of greater therapeutic potential than inhibition of P_{2Y2}-mediated mucin secretion [4], and there is considerable interest in development of P_{2Y2} agonists. In phase I clinical trial, a second generation P_{2Y2} agonist, INS365, was safe, well tolerated and significantly enhanced sputum expectoration [87]. The compound is now in phase II trials.

Macrolide antibiotics

Antibiotic therapy for acute infections is recommended in the management of COPD [2]. Erythromycin is a macrolide antibiotic that inhibits mucin secretion in a variety of experimental preparations, including human airways [10], and anecdotally reduces excessive mucus secretion in patients [89]. The mechanism of action of erythromycin is unclear, but may involve anti-inflammatory rather than mucoactive effects [90, 91]. Formal clinical studies of its effects on the pathophysiology of mucus hypersecretion (i.e., sputum production and lung function) in COPD would be of interest.

Conclusions and future directions

Airway mucus hypersecretion and the pathophysiological changes that accompany it (e.g., goblet cell hyperplasia) are inconsistent features in COPD. The impact on morbidity and mortality is limited to certain groups of patients, particularly those that are prone to respiratory tract infection. Nevertheless, it is important to develop drugs that inhibit mucus hypersecretion in these patients. Before these issues can be addressed, considerably more information is needed concerning the biochemical and biophysical nature of airway mucins in normal healthy subjects, whether or not there is an intrinsic abnormality of mucus in COPD, and whether any abnormality is specific for COPD. In addition, the factors that regulate MUC gene expression in health and disease, and the relationship between this regulation and the development of the hypersecretory phenotype, need to be determined. The above information can be used in optimising identification of therapeutic targets which, in turn, should lead to rational design of anti-hypersecretory drugs which may have to be specific for COPD.

References

1. Fletcher CM, Pride NB (1984) Definitions of emphysema, chronic bronchitis, asthma, and airflow obstruction: 25 years on from the Ciba symposium. *Thorax* 39: 81–85
2. National Heart Lung and Blood Institute, WHO (2001) Global initiative for chronic obstructive lung disease. National Institutes of Health, publication number 2701
3. Prescott E, Lange P, Vestbo J (1995) Chronic mucus hypersecretion in COPD and death from pulmonary infection. *Eur Respir J* 8: 1333–1338
4. Knowles MR, Boucher RC (2002) Mucus clearance as a primary innate defense mechanism for mammalian airways. *J Clin Invest* 109: 571–577
5. Davies JR, Herrmann A, Russell W, Svitacheva N, Wickström C, Carlstedt I (2002) Respiratory tract mucins: structure and expression patterns. In: Novartis Foundation Symposium: *Mucus Hypersecretion in Respiratory Disease*. John Wiley & Sons, Chichester, 76–88
6. Dekker J, Rossen JW, Buller HA, Einerhand AW (2002) The MUC family: an obituary. *Trends Biochem Sci* 27: 126–131
7. Lapensee L, Paquette Y, Bleau G (1997) Allelic polymorphism and chromosomal localization of the human oviductin gene (MUC9). *Fertil Steril* 68: 702–708
8. Gum JR, Jr., Crawley SC, Hicks JW, Szymkowski DE, Kim YS (2002) MUC17, a novel membrane-tethered mucin. *Biochem Biophys Res Commun* 291: 466–475
9. Wu GJ, Wu MW, Wang SW, Liu Z, Qu P, Peng Q, Yang H, Varma VA, Sun QC, Petros JA et al (2001) Isolation and characterization of the major form of human MUC18 cDNA gene and correlation of MUC18 over-expression in prostate cancer cell lines and tissues with malignant progression. *Gene* 279: 17–31
10. Rogers DF (2000) Mucus pathophysiology in COPD: differences to asthma, and pharmacotherapy. *Monaldi Arch Chest Dis* 55: 324–332
11. Reid L (1954) Pathology of chronic bronchitis. *Lancet* i: 275–278
12. Aikawa T, Shimura S, Sasaki H, Takishima T, Yaegashi H, Takahashi T (1989) Morphometric analysis of intraluminal mucus in airways in chronic obstructive pulmonary disease. *Am Rev Respir Dis* 140: 477–482
13. Steiger D, Fahy J, Boushey H, Finkbeiner WE, Basbaum C (1994) Use of mucin antibodies and cDNA probes to quantify hypersecretion *in vivo* in human airways. *Am J Respir Cell Mol Biol* 10: 538–545
14. Mullen JB, Wright JL, Wiggs BR, Pare PD, Hogg JC (1987) Structure of central airways in current smokers and ex-smokers with and without mucus hypersecretion: relationship to lung function. *Thorax* 42: 843–848
15. Reid L (1960) Measurement of the bronchial mucous gland layer: a diagnostis yardstick in chronic bronchitis. *Thorax* 15: 132–141
16. Restrepo G, Heard BE (1963) The size of the bronchial glands in chronic bronchitis. *J Pathol Bacteriol* 85: 305–310
17. Glynn AA, Michaels L (1960) Bronchial biopsy in chronic bronchitis and asthma. *Thorax* 15: 142–153

18   Hayes JA (1960) Distribution of bronchial gland measurement in a Jamaican population. *Thorax* 24: 619–622
19   Thurlbeck WM, Angus CW, Paré JAP (1963) Mucous gland hypertrophy in chronic bronchitis, and its occurence in smokers. *Brit J Dis Chest* 57: 73–78
20   Thurlbeck WM, Angus GE (1964) A distribution curve for chronic bronchitis. *Thorax* 19: 436–442
21   Davies JR, Hovenberg HW, Linden CJ, Howard R, Richardson PS, Sheehan JK, Carlstedt I (1996) Mucins in airway secretions from healthy and chronic bronchitic subjects. *Biochem J* 313: 431–439
22   Thornton DJ, Carlstedt I, Howard M, Devine PL, Price MR, Sheehan JK (1996) Respiratory mucins: identification of core proteins and glycoforms. *Biochem J* 316: 967–975
23   Hovenberg HW, Davies JR, Herrmann A, Linden CJ, Carlstedt I (1996) MUC5AC, but not MUC2, is a prominent mucin in respiratory secretions. *Glycoconj J* 13: 839–847
24   Wickstrom C, Davies JR, Eriksen GV, Veerman EC, Carlstedt I (1998) MUC5B is a major gel-forming, oligomeric mucin from human salivary gland, respiratory tract and endocervix: identification of glycoforms and C-terminal cleavage. *Biochem J* 334: 685–693
25   Sheehan JK, Howard M, Richardson PS, Longwill T, Thornton DJ (1999) Physical characterization of a low-charge glycoform of the MUC5B mucin comprising the gel-phase of an asthmatic respiratory mucous plug. *Biochem J* 338: 507–513
26   Kirkham S, Sheehan JK, Knight D, Richardson PS, Thornton DJ (2002) Heterogeneity of airways mucus: variations in the amounts and glycoforms of the major oligomeric mucins MUC5AC and MUC5B. *Biochem J* 361: 537–546
27   Chen Y, Zhao YH, Di YP, Wu R (2001) Characterization of human mucin 5B gene expression in airway epithelium and the genomic clone of the amino-terminal and 5'-flanking region. *Am J Respir Cell Mol Biol* 25: 542–553
28   Groneberg DA, Eynott PR, Lim S, Oates T, Wu R, Carlstedt I, Roberts P, McCann B, Nicholson AG, Harrison BD, Chung KF (2002) Expression of respiratory mucins in fatal status asthmaticus and mild asthma. *Histopathology* 40: 367–373
29   Groneberg DA, Eynott PR, Oates T, Lin S, Wu R, Carlstedt I, Nicholson AG, Chung KF (2002) Expression of MUC5AC and MUC5B mucins in normal and cystic fibrosis lung. *Respir Med* 96: 81–86
30   Davies JR, Svitacheva N, Lannefors L, Kornfalt R, Carlstedt I (1999) Identification of MUC5B, MUC5AC and small amounts of MUC2 mucins in cystic fibrosis airway secretions. *Biochem J* 344 Pt 2: 321–330
31   Wanner A (1977) Clinical aspects of mucociliary transport. *Am Rev Respir Dis* 116: 73–125
32   McDowell EM, Barrett LA, Harris CC, Trump BF (1976) Abnormal cilia in human bronchial epithelium. *Arch Pathol Lab Med* 100: 429–436
33   Wanner A, Salathe M, O'Riordan TG (1996) Mucociliary clearance in the airways. *Am J Respir Crit Care Med* 154: 1868–1902
34   Pavia D, Agnew JE, Glassman JM, Sutton PP, Lopez-Vidriero MT, Soyka JP, Clarke SW

(1985) Effects of iodopropylidene glycerol on tracheobronchial clearance in stable, chronic bronchitic patients. *Eur J Respir Dis* 67: 177–184

35 Moretti M, Lopez-Vidriero MT, Pavia D, Clarke SW (1997) Relationship between bronchial reversibility and tracheobronchial clearance in patients with chronic bronchitis. *Thorax* 52: 176–180

36 Goodman RM, Yergin BM, Landa JF, Golivanux MH, Sackner MA (1978) Relationship of smoking history and pulmonary function tests to tracheal mucous velocity in non-smokers, young smokers, ex-smokers, and patients with chronic bronchitis. *Am Rev Respir Dis* 117: 205–214

37 Agnew JE, Little F, Pavia D, Clarke SW (1982) Mucus clearance from the airways in chronic bronchitis – Smokers and ex-smokers. *Bull Eur Physiopathol Respir* 18: 473–484

38 van der Schans CP, Piers DA, Beekhuis H, Koeter GH, van der Mark TW, Postma DS (1990) Effect of forced expirations on mucus clearance in patients with chronic airflow obstruction: effect of lung recoil pressure. *Thorax* 45: 623–627

39 Ericsson CH, Svartengren K, Svartengren M, Mossberg B, Philipson K, Blomquist M, Canner P (1995) Repeatability of airway deposition and tracheobronchial clearance rate over three days in chronic bronchitis. *Eur Respir J* 8: 1886–1893

40 Fletcher C, Peto R (1977) The natural history of chronic airflow obstruction. *Br Med J* 1: 1645–1648

41 Kauffmann F, Drouet D, Lellouch J, Brille D (1979) Twelve years spirometric changes among Paris area workers. *Int J Epidemiol* 8: 201–212

42 Higgins MW, Keller JB, Becker M, Howatt W, Landis JR, Rotmann H, Weg JG, Higgins I (1982) An index of risk for obstructive airways disease. *Am Rev Respir Dis* 125: 144–151

43 Peto R, Speizer FE, Cochrane AL, Moore F, Fletcher CM, Tinker CM, Higgins JT, Gray RG, Richards SM, Gilliland J, Norman-Smith B (1983) The relevance in adults of airflow obstruction, but not of mucus hypersecretion, to mortality from chronic lung disease. Results from 20 years of prospective observation. *Am Rev Respir Dis* 128: 491–500

44 Ebi-Kryston KL (1988) Respiratory symptoms and pulmonary function as predictors of 10-year mortality from respiratory disease, cardiovascular disease, and all causes in the Whitehall Study. *J Clin Epidemiol* 41: 251–260

45 Annesi I, Kauffmann F (1986) Is respiratory mucus hypersecretion really an innocent disorder? A 22-year mortality survey of 1,061 working men. *Am Rev Respir Dis* 134: 688–693

46 Vestbo J, Prescott E, Lange P (1996) Association of chronic mucus hypersecretion with FEV1 decline and chronic obstructive pulmonary disease morbidity. Copenhagen City Heart Study Group. *Am J Respir Crit Care Med* 153: 1530–1535

47 Lange P, Nyboe J, Appleyard M, Jensen G, Schnohr P (1990) Relation of ventilatory impairment and of chronic mucus hypersecretion to mortality from obstructive lung disease and from all causes. *Thorax* 45: 579–585

48 Speizer FE, Fay ME, Dockery DW, Ferris BG, Jr. (1989) Chronic obstructive pulmonary disease mortality in six U.S. cities. *Am Rev Respir Dis* 140: S49–S55
49 Sherman CB, Xu X, Speizer FE, Ferris BG Jr, Weiss ST, Dockery DW (1992) Longitudinal lung function decline in subjects with respiratory symptoms. *Am Rev Respir Dis* 146: 855–859
50 Rogers DF (1994) Airway goblet cells: responsive and adaptable front-line defenders. *Eur Respir J* 7: 1690–1706
51 Tamaoki J, Chiyotani A, Kobayashi K, Sakai N, Kanemura T, Takizawa T (1992) Effect of indomethacin on bronchorrhea in patients with chronic bronchitis, diffuse panbronchiolitis, or bronchiectasis. *Am Rev Respir Dis* 145: 548–552
52 Zeldin DC (2001) Epoxygenase pathways of arachidonic acid metabolism. *J Biol Chem* 276: 36059–36062
53 Tesfaigzi Y, Kluger M, Kozak W (2001) Clinical and cellular effects of cytochrome P-450 modulators. *Respir Physiol* 128: 79–87
54 Liu YC, Khawaja AM, Rogers DF (1998) Pathophysiogy of Airway Mucus Secretion in Asthma. In: PJ Barnes, IW Rodger, NC Thomson (eds): *Asthma. Basic Mechanisms and Clinical Management.* Academic Press, London, 205–227
55 Nadel JA, Burgel PR (2001) The role of epidermal growth factor in mucus production. *Curr Opin Pharmacol* 1: 254–258
56 Sturton G, Fitzgerald M (2002) Phosphodiesterase 4 inhibitors for the treatment of COPD. *Chest* 121: 192S–196S
57 Nadel JA, Takeyama K, Agusti C (1999) Role of neutrophil elastase in hypersecretion in asthma. *Eur Respir J* 13: 190–196
58 Witko-Sarsat V, Halbwachs-Mecarelli L, Schuster A, Nusbaum P, Ueki I, Canteloup S, Lenoir G, Descamps-Latscha B, Nadel JA (1999) Proteinase 3, a potent secretagogue in airways, is present in cystic fibrosis sputum. *Am J Respir Cell Mol Biol* 20: 729–736
59 Nadel JA (2000) Role of neutrophil elastase in hypersecretion during COPD exacerbations, and proposed therapies. *Chest* 117: 386S–389S
60 Nogami H, Aizawa H, Matsumoto K, Nakano H, Koto H, Miyazaki H, Hirose T, Nishima S, Hara N (2000) Neutrophil elastase inhibitor, ONO-5046 suppresses ozone-induced airway mucus hypersecretion in guinea pigs. *Eur J Pharmacol* 390: 197–202
61 Gorrini M, Lupi A, Viglio S, Pamparana F, Cetta G, Iadarola P, Powers JC, Luisetti M (2001) Inhibition of human neutrophil elastase by erythromycin and flurythromycin, two macrolide antibiotics. *Am J Respir Cell Mol Biol* 25: 492–499
62 MacNee W (2001) Oxidative stress and lung inflammation in airways disease. *Eur J Pharmacol* 429: 195–207
63 Montuschi P, Kharitonov SA, Barnes PJ (2001) Exhaled carbon monoxide and nitric oxide in COPD. *Chest* 120: 496–501
64 Wright DT, Fischer BM, Li C, Rochelle LG, Akley NJ, Adler KB (1996) Oxidant stress stimulates mucin secretion and PLC in airway epithelium *via* a nitric oxide-dependent mechanism. *Am J Physiol* 271: L854–L861

65 Rogers DF (2002) Mucoactive drugs for asthma and COPD: any place in therapy? *Expert Opin Investig Drugs* 11: 15–35

66 Rogers DF (2002) Pharmacological regulation of the neuronal control of airway mucus secretion. *Curr Opin Pharmacol* 2: 249–255

67 Ghafouri MA, Patil KD, Kass I (1984) Sputum changes associated with the use of ipratropium bromide. *Chest* 86: 387–393

68 Tamaoki J, Chiyotani A, Tagaya E, Sakai N, Konno K (1994) Effect of long term-treatment with oxitropium bromide on airway secretion in chronic bronchitis and diffuse panbronchiolitis. *Thorax* 49: 545–548

69 Disse B (2001) Antimuscarinic treatment for lung diseases from research to clinical practice. *Life Sci* 68: 2557–2564

70 Mitsuya M, Mase T, Tsuchiya Y, Kawakami K, Hattori H, Kobayashi K, Ogino Y, Fujikawa T, Satoh A, Kimura T et al (1999) J-104129, a novel muscarinic M3 receptor antagonist with high selectivity for M3 over M2 receptors. *Bioorg Med Chem* 7: 2555–2567

71 Rogers DF (2001) Tachykinin receptor antagonists for asthma and COPD. *Expert Opin Ther Patents* 11: 1097–1121

72 Spicuzza L, Haddad EB, Birrell M, Ling A, Clarke D, Venkatesan P, Barnes PJ, Belvisi MG (2000) Characterization of the effects of cannabinoids on guinea-pig tracheal smooth muscle tone: role in the modulation of acetylcholine release from parasympathetic nerves. *Br J Pharmacol* 130: 1720–1726

73 Caterina MJ, Julius D (2001) The vanilloid receptor: a molecular gateway to the pain pathway. *Annu Rev Neurosci* 24: 487–517

74 Pertwee RG (2001) Cannabinoid receptors and pain. *Prog Neurobiol* 63: 569–611

75 Li Y, Martin LD, Spizz G, Adler KB (2001) MARCKS protein is a key molecule regulating mucin secretion by human airway epithelial cells *in vitro*. *J Biol Chem* 276: 40982–40990

76 Evans C, Kheradmand F, Corry D, Tuvim M, Densmore C, Waldrep C, Knight V, Dickey B (2002) Gene therapy of mucus hypersecretion in experimental asthma. *Chest* 121: 90S–91S

77 Wang B, Lim DJ, Han J, Kim YS, Basbaum CB, Li JD (2002) Novel cytoplasmic proteins of nontypeable *Haemophilus influenzae* up-regulate human MUC5AC mucin transcription *via* a positive p38 mitogen-activated protein kinase pathway and a negative phosphoinositide 3-kinase-Akt pathway. *J Biol Chem* 277: 949–957

78 Shim JJ, Dabbagh K, Takeyama K, Burgel PR, Dao-Pick TP, Ueki IF, Nadel JA (2000) Suplatast tosilate inhibits goblet-cell metaplasia of airway epithelium in sensitized mice. *J Allergy Clin Immunol* 105: 739–745

79 Longphre M, Li D, Li J, Matovinovic E, Gallup M, Sarnet JM, Basbaum CB (2000) Lung mucin production is stimulated by the air pollutant residual oil fly ash. *Toxicol Appl Pharmacol* 162: 86–92

80 Nakanishi A, Morita S, Iwashita H, Sagiya Y, Ashida Y, Shirafuji H, Fujisawa Y,

Nishimura O, Fujino M (2001) Role of gob-5 in mucus overproduction and airway hyperresponsiveness in asthma. *Proc Natl Acad Sci USA* 98: 5175–5180

81 Tesfaigzi Y, Fischer MJ, Martin AJ, Seagrave J (2000) Bcl-2 in LPS- and allergen-induced hyperplastic mucous cells in airway epithelia of Brown Norway rats. *Am J Physiol Lung Cell Mol Physiol* 279: L1210–L1217

82 Tesfaigzi Y, Fischer MJ, Daheshia M, Green FH, De Sanctis GT, Wilder JA (2002) Bax is crucial for IFN-gamma-induced resolution of allergen-induced mucus cell metaplasia. *J Immunol* 169: 5919–5925

83 Bhattacharyya SN, Manna B, Smiley R, Ashbaugh P, Coutinho R, Kaufman B (1998) Smoke-induced inhalation injury: effects of retinoic acid and antisense oligodeoxynucleotide on stability and differentiated state of the mucociliary epithelium. *Inflammation* 22: 203–214

84 Poole PJ, Black PN (2000) Mucolytic agents for chronic bronchitis or chronic obstructive pulmonary disease. *Cochrane Database Syst Rev* CD001287

85 Gillissen A, Jaworska M, Orth M, Coffiner M, Maes P, App EM, Cantin AM, Schultze-Werninghaus G (1997) Nacystelyn, a novel lysine salt of N-acetylcysteine, to augment cellular antioxidant defence *in vitro*. *Respir Med* 91: 159–168

86 App EM, Baran D, Dab I, Malfroot A, Coffiner M, Vanderbist F, King M (2002) Dose-finding and 24-h monitoring for efficacy and safety of aerosolized Nacystelyn in cystic fibrosis. *Eur Respir J* 19: 294–302

87 Kellerman DJ (2002) P2Y(2) receptor agonists: a new class of medication targeted at improved mucociliary clearance. *Chest* 121: 201S–205S

88 Roger P, Gascard JP, Bara J, de Montpreville VT, Yeadon M, Brink C (2000) ATP induced MUC5AC release from human airways *in vitro*. *Mediators Inflamm* 9: 277–284

89 Marom ZM, Goswami SK (1991) Respiratory mucus hypersecretion (bronchorrhea): a case discussion – possible mechanisms(s) and treatment. *J Allergy Clin Immunol* 87: 1050–1055

90 Wales D, Woodhead M (1999) The anti-inflammatory effects of macrolides. *Thorax* 54 (Suppl 2): S58–S62

91 Shibuya Y, Wills PJ, Cole PJ (2001) The effect of erythromycin on mucociliary transportability and rheology of cystic fibrosis and bronchiectasis sputum. *Respiration* 68: 615–619

Induced sputum and BAL analysis in COPD

Vera M. Keatings[1] and Clare M. O'Connor[2]

[1]Department of Respiratory Medicine, Letterkenny General Hospital, Letterkenny, Co. Donegal, Ireland; [2]The Conway Institute for Biomolecular and Biomedical Research and the Dublin Molecular Medical Centre, University College Dublin, Belfield, Dublin 4, Ireland

Introduction

An essential component of airways inflammation research is the examination of airway luminal fluid. Since its development as a sampling technique over 20 years ago, bronchoalveolar lavage (BAL) has become the most widely used method of evaluation of distal airway and alveolar inflammation. The widespread use of BAL reflects its relative ease and lack of clinical complications, particularly in comparison to airway or lung biopsy. However, while BAL is now a universally used tool in research and diagnosis of interstitial lung diseases (ILDs) its use in evaluating airway inflammation in COPD is less common. The lesser use of BAL in COPD studies reflects a lower level of tolerance of the procedure in patients with moderate to severe airway obstruction and greater difficulty in obtaining adequate samples from these same patient groups [1, 2].

In recent years the induction of sputum using normal or hypertonic saline has been developed as a safe valid method of harvesting luminal fluid [3]. Its composition does not differ inherently from spontaneously produced sputum but it contains a higher proportion of viable cells, facilitating analysis [4]. Unlike spontaneous sputum, it can be obtained from all patients, even those who do not produce sputum, and enables sampling of control subjects. To date it has been most widely used in studies on asthma, but has also been employed to evaluate a range of inflammatory mediators in COPD.

The focus of this chapter is to review the information on airway inflammation obtained by both procedures to provide a basis for selection of sampling method for research studies. Information obtained using induced sputum will be evaluated first, followed by that from BAL studies and the two compared.

Induced sputum as a method for the study of airway inflammation

COPD has a long pre-clinical course. Usually, moderate airflow obstruction is already established at the time of clinical presentation and invasive procedures for research purposes are relatively contraindicated. A non-invasive method of obtaining luminal fluid is therefore advantageous.

Induction of sputum using inhaled nebulised hypertonic saline was first used to obtain bronchopulmonary secretions from patients with suspected Pneumocystis carinii pneumonia in the late 1980's [5] from patients who were unable to produce sputum spontaneously and continues to be a useful tool for this purpose. This led to interest in the technique as a method of obtaining airway cells and fluid for research purposes [6]. Sputum induction has now been extensively used to study airways disease. It is an effective inexpensive method of demonstrating airway inflammation. Its safety has been established thus allowing research involving patients with moderate or severe disease [7]. Uniquely, the kinetics of the inflammatory response can be studied, as the sputum induction can be repeated.

Sputum is induced by the inhalation of saline in concentrations varying from 0.9% up to 7%. Different protocols exist regarding both for the induction method and the processing of the resultant expectorate. The method of sputum induction has little effect on the cellular findings [8, 9]. Due to the variety of methods in use, an ERS taskforce recently reached a consensus that 4.5% saline should be used as standard [10]. Two methods of sputum processing have emerged, one using the entire sample of expectorate [11] the other selecting plugs of sputum using an inverted microscope [12]. Both methods have advantages and disadvantages but are not interchangeable. It is therefore recommended that findings from studies using these different methods of sputum processing should not be directly compared [13].

Cellular findings in induced sputum in COPD

Induced sputum from patients with COPD contains markedly more cells than that from control subjects and neutrophil counts are significantly higher (Fig 1). The degree of neutrophilia has been shown to be inversely proportional to the FEV_1, patients with more severe disease having a greater neutrophilia. This finding has been interpreted variably, most suggesting that it indicates a pivotal role for the neutrophil in the pathogenesis of airflow limitation, others suggesting a mechanical effect of a narrowed airway resulting in a bias of the type of inflammatory cells expectorated. However, smokers who have not developed lung disease also have a significant number of neutrophils in their airways compared with non-smokers [14]. Therefore it is vital, in order to determine the effect of smoking *versus* a disease effect, to use control subjects who have a similar smoking history.

Studies examining neutrophil granule proteins such as myeloperoxidase and human neutrophil lipocalin (HNL) demonstrate markedly raised concentrations of these products indicating neutrophil activation in COPD [11]. When markers of eosinophilic inflammation in COPD are examined, some series of patients have been shown to have either increased eosinophil percentages, increased absolute num-

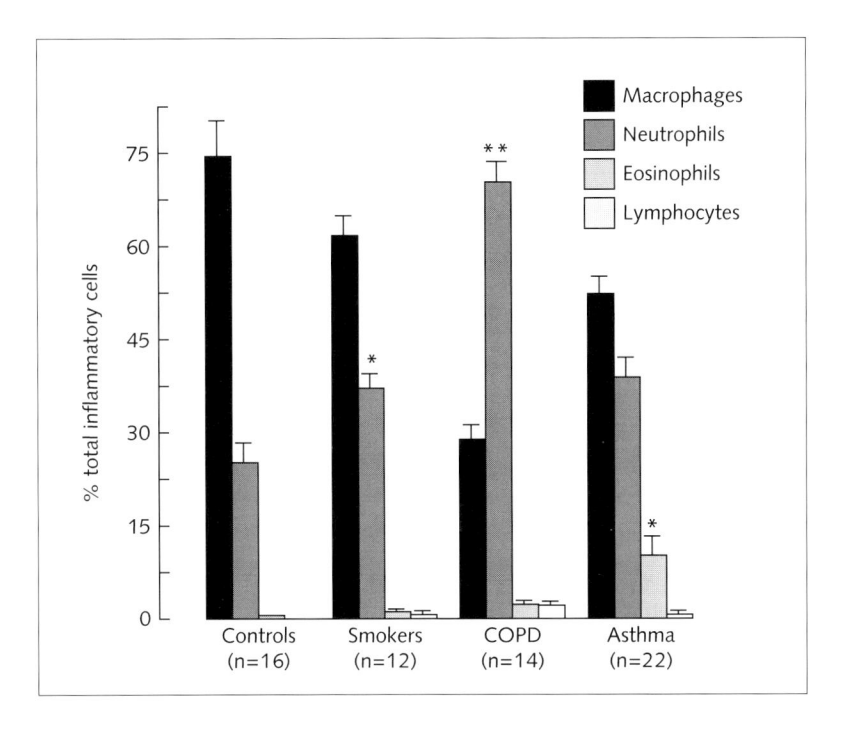

Figure 1
Differential cell counts in induced sputum in subject groups. Cell counts are expressed as percentages of total inflammatory cells. Data are means. Vertical bars denote SEM. *p < 0.05 compared to controls, **p < 0.001 compared to all other groups. Reproduced from [14].

bers/g of sputum associated with increased concentrations of the granule proteins concentrations eosinophil cationic protein (ECP) and eosinophil peroxidase (EPO) [11, 15]. It has been suggested that the raised eosinophil granule protein concentrations are a non-specific finding but similar concentrations of ECP in asthma are considered to be implicated in the disease process and considered useful markers of disease severity [3]. Some studies have identified patients with COPD who have a significant eosinophilia in induced sputum and have attempted to define the significance of this. Louis et al. [16] found comparable concentrations of ECP in asthma and COPD and furthermore found that those with eosinophilia had raised concentrations of tryptase, which has been characterized as a hallmark of asthmatic inflammation for many years. Equally, they could not discriminate between the two conditions by measurement of either tryptase or sICAM-1. Intuitively one might expect that eosinophilic inflammation may be found in patients with more asthmatic features but when the clinical characteristics of "eosinophilic COPD" (eosinophils

>3%) were compared with those patients with "non-eosinophilic" COPD, no differences could be found in terms of age, FEV$_1$, reversibility of FEV$_1$ or diffusing capacity (DLCO).

Soluble mediators

To further examine airway inflammation in COPD, inflammatory cytokines have been analysed in sputum in a range of studies. Two methods are used to obtain sputum fluid. The "sol" phase of sputum, obtained by ultracentrifugation of untreated sputum at 60,000 g for 90 minutes is a clear low volume supernatant, separated from a thick pellet of mucus and cells. In order to prepare supernatant more quickly, many investigators disperse the sputum using the reducing agent dithiothreitol (DTT), which reduces the disulphide bonds of mucin allowing separation of fluid by centrifugation at 300–400 g for 10 minutes. Where this method is used confirmation that DTT does not interfere with either the assay process or destroy the cytokine is essential [17]. This is more than a theoretical point as many cytokines e.g., IL-8 contain disulphide bonds. In addition some analytes may be broken down by proteases which are present in the sputum and protease or peptidase inhibitors may need to be added to the supernatant prior to storage [18]. All assays should be validated in the biological fluid in which the measurements are to take place and standard curves altered accordingly [17].

To date most studies have examined the role of mediators involved in neutrophilic inflammation in COPD. IL-8, a CXC chemokine that acts as a potent neutrophil chemoattractant, is present in high concentrations in induced sputum from patients with COPD compared to controls and is believed to play an important part in propagation of the neutrophilic inflammation in this condition [14]. It is also raised in the sputum of healthy smokers and may play a part in the initiation of neutrophilic inflammation [14]. LTB$_4$, another potent neutrophil chemoattractant, is also present in sputum from patients with COPD in concentrations proportional to the degree of neutrophilic inflammation and it too is considered to play an important part in neutrophil chemotaxis [19]. Growth related oncogene-α (GRO-α) is another CXC chemokine which is powerfully chemotactic for neutrophils which is present in markedly raised concentrations in patients with COPD compared with controls and healthy smokers [20]. IL-8 and GRO-α production by alveolar macrophages is enhanced by exposure to cigarette smoke [21].

Another macrophage-produced cytokine, TNF-α, is raised in patients with COPD compared with both non-smoking controls and smokers and therefore is more specific for the presence of COPD than IL-8 (Fig. 2). When administered by inhalation to the normal lung TNF-α has been shown to result in an influx of neutrophils [22], possibly *via* activation of transcription factor NF-κB and so increasing transcription of the IL-8 gene. Given the increase in macrophage-derived

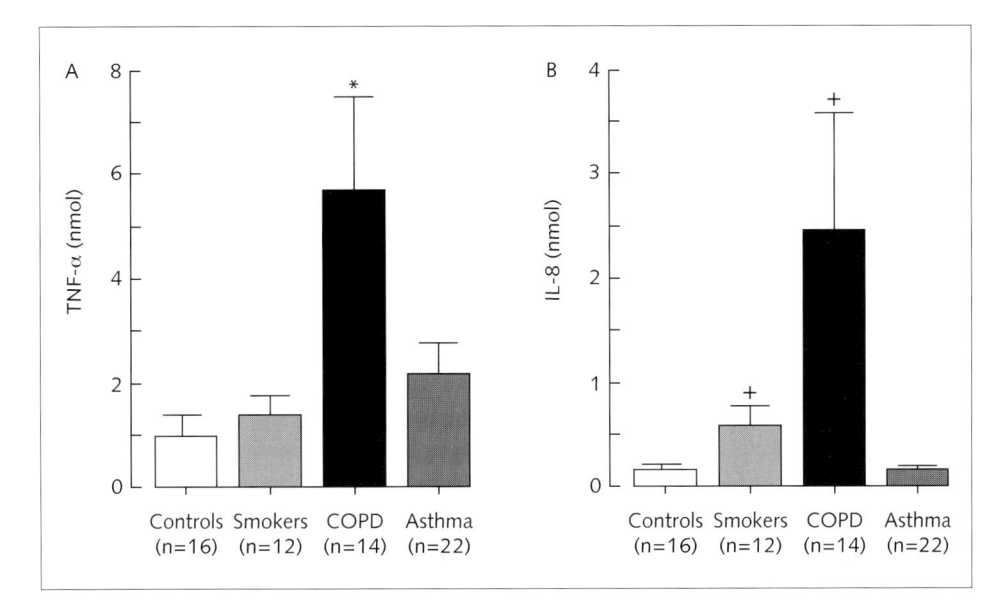

Figure 2
Concentration of TNF-α (A) and IL-8 (B) in induced sputum in subjects with asthma, COPD, smokers and non-smoking control subjects. Data are means. Vertical bars denote SEM. *p<0.05 compared with smokers and p<0.01 compared with non-smoking controls. +p<0.05 compared with non-smoking controls. Reproduced from [14].

cytokines, it is of interest that the macrophage chemotactic protein-1 (MCP-1) is also present in markedly raised concentrations in induced sputum of patients with COPD, although there is a wide variation in concentration between individuals reflecting the heterogeneity of patients with COPD [20].

IL-10, an anti-inflammatory cytokine which down-regulates TNF-α production and gene expression in alveolar macrophages [23] has been shown to be reduced in induced sputum from COPD patients with immunocytochemistry localizing this cytokine to the alveolar macrophages and lymphocytes [24]. It might therefore be suggested that failure of alveolar macrophages to produce sufficient immunoregulatory cytokines such as IL-10 may contribute to the excessive inflammation seen in the respiratory tract in COPD.

In summary, the cytokine and mediator profile seen in COPD differs significantly from that seen in asthma. In COPD there is an infiltrate of neutrophils and CD8+ lymphocytes. The cytokine TNF-α and neutrophil chemoattractants IL-8, GRO-α and LTB$_4$ appear to have a prominent role. There is redundancy in that many cytokines may perform the same role and similarly one cytokine may be produced by a number of different cells and number of effects on a wide variety of cells. Measurement of soluble mediators in the supernatant of induced sputum is therefore useful as a guide to the importance of individual mediators but should be complementary to biopsies, immunocytochemistry and functional studies.

Proteases and antiproteases

An increase in elastase activity relative to the lung's anti-elastase screen has for long been implicated in alveolar tissue destruction in COPD. More recently, the involvement of matrix metalloproteinases (MMPs) and their inhibitors (tissue inhibitors of metalloproteinases (TIMPs)) has also been implicated in both the tissue destruction and airway remodelling components of the disease [25, 26]. Active, unopposed elastase has been detected in induced sputum (IS) from patients with chronic bronchitis and from stable COPD patients [27, 28], suggesting that elastase-anti-elastase imbalance can be detected by IS sampling of the airways. Vignola et al. [27] also found that IS elastase was inversely proportional to degree of airway obstruction and to percentage of sputum neutrophils. Studies evaluating MMP/TIMP balance in the airways by IS have reported an increase in MMP-9 (gelatinase B) in patients with COPD and chronic bronchitis [26, 29] and the MMP-9/TIMP-1 ratio was correlated with FEV_1, suggesting a decrease in function as the balance in the airways tips in favour of the enzyme rather than the inhibitor [26]. Although these studies are encouraging, further work is necessary to fully appraise the value of IS in assessing protease/antiprotease imbalance in the airways.

Relationship between sputum inflammation and disease phenotype

In clinical practice, COPD is a heterogeneous condition with individual patients having varying clinical characteristics in terms of reversibility, cachexia, type 1 or 2 respiratory failure sputum production, the presence of emphysema, the development of *cor pulmonale* and the frequency of exacerbations. This definition clearly includes patients that differ enormously and may have little else in common other than that they don't have asthma. While the original studies using induced sputum allowed differentiation between asthma and COPD, investigators are now attempting to differentiate between patient groups within COPD on the basis of airway inflammation. Such studies will hopefully allow patients within COPD to be better defined facilitating more focused therapeutic strategies. The following section describes studies that related airway inflammation to the clinical characteristics of exacerbation frequency and steroid responsiveness.

Two major studies have compared the inflammation in sputum between disease with frequent exacerbations and that with infrequent exacerbations. One study found that patients who exacerbate more frequently (> 3/y) have higher concentrations of

IL-6 and IL-8 in induced sputum than patients who had ≤2 exacerbations per year [30]. In apparent contradiction, a study by Gompertz et al. [31] found no difference in IL-8 or LTB$_4$ concentrations at baseline between frequent and infrequent exacerbators. The discordance between these studies may be explained by differences in patient phenotype for example, the patients in the study of Gompertz et al. all produced spontaneous sputum, while this is not necessarily true for those in Bhowmik's study [30]. An important difference in patients phentoype was introduced because Gompertz et al. excluded a number of patients with tubular bronchiectasis on high resolution CT (HRCT) scan who otherwise would have fulfilled the definition of COPD for the study by Bhowmik and most other studies published on COPD. The apparent disparity between these studies underlines the heterogeneity of COPD and the need for precise definition of phenotype in clinical studies.

A study examining the effect of short-term oral steroid therapy on soluble mediators have shown no reduction in concentrations of IL-8, TNF-α, ECP, MPO, HNL after two weeks of high dose oral prednisolone or inhaled budesonide [32]. By comparison with asthma patients, patients with COPD had lower numbers of eosinophils and no response to steroids either in terms of FEV$_1$, eosinophil numbers or granule proteins. Pizzichini et al. [33] further examined the relationship between the presence of a sputum eosinophilia and short-term response to oral steroids. Patients with sputum eosinophilia were similar to the non-eosinophilic subjects in terms of spirometry, reversibility to β$_2$ agonist, smoking history and atopy. The patients with eosinophilia had a small but significant increase in FEV$_1$ (mean 0.1 L) and a significant improvement in quality of life scores compared with response to placebo, while patients without eosinophilia had no response in either quality of life or lung function. These findings were confirmed by a larger double-blind placebo controlled study demonstrating that patients with sputum eosinophilia had improvements in terms of FEV$_1$ and QOL [34]. While the improvement in FEV$_1$ was still small enough to consider the airflow obstruction "irreversible", these studies nevertheless demonstrate that sputum eosinophilia in COPD is not simply a transient phenomenon, and that eosinophils have significant functional effects within the airways. This study could be interpreted simply as demonstrating that the eosinophilic patients had asthma in addition to COPD but crucially this is not detectable from the patients clinical characteristics. Sputum inflammatory cell/mediator measurement may prove an important indicator in treatment planning. The relationship between sputum markers and the long-term response to inhaled steroids has not yet been examined.

BAL analysis of airway inflammation in COPD

As mentioned in the introductory section, BAL is not well tolerated in COPD patients with moderate to severe disease, hence information on airway inflammation obtained from BAL studies is largely confined to patients with mild to moderate disease. Even in these patients, the direct correlation observed between degree of airflow obstruction and yield of BAL sample [2] underlines an inherent problem associated with this technique in evaluating airway inflammation in COPD. In the absence of rigorous normalisation for sample yield, erroneous conclusions on the relationship between cellular and soluble analytes and airway obstruction can easily be made.

Despite these drawbacks, significant information on the inflammatory process in COPD has been obtained from BAL studies and, if induced sputum is to be used instead of BAL in evaluating airway inflammation in this disease, it is necessary to ensure that it provides at least as much information. The remainder of this section will concentrate on the COPD inflammatory profile revealed by BAL studies with a focus on whether or not induced sputum studies are similarly informative.

BAL evaluation of cellular components of airway inflammation in COPD

Increased cellularity in BAL from patients with COPD compared to healthy controls primarily reflects cellular infiltrate resulting from smoking, as little change in cell numbers recovered by BAL between diseased and non-diseased smokers or ex-smokers is observed [2, 35, 36]. Indeed, a reduction in inflammatory cell numbers with increased airway obstruction has been reported, a decrease that is not wholly accounted for by the lower recovery of lavage fluid [37].

In respect of cellular composition, studies have variously implicated neutrophils, macrophages, CD8+ve lymphocytes and eosinophils as important effector cells in the inflammatory process in COPD. Significant support for the involvement of the neutrophil came from early studies which reported increased neutrophils in BAL from patients with chronic bronchitis and emphysema [1, 38], with particularly elevated levels being observed in patients with alpha-one proteinase inhibitor (α_1PI) deficiency emphysema [39]. However, in later studies where the majority of patients had mild to moderate disease, increased BAL neutrophils is not commonly observed [25, 36]. This discrepancy likely reflects an increase in airway neutrophils as disease progresses, as elevated BAL neutrophils has been shown to be associated with increased air-flow obstruction [40]. Similar associations between increased neutrophils and disease severity have been observed in tissue and induced sputum samples [14, 41]. This suggests that airway neutrophilia may be more useful in evaluating disease progression than in determining the presence or absence of disease. In this context, evaluating neutrophil infiltration by induced sputum is most suited for the repeat measurements required for progression studies. Indeed, a study by Martin et al. [38], where each BAL aliquot was analysed separately, suggests that increased BAL neutrophils may derive primarily from the upper airways, the site sampled by induced sputum.

Although an increase in BAL neutrophils is not a consistent finding in mild to moderate disease, neutrophil activation, as reflected by the release of the granule marker HNL and granule proteases (discussed below), may be a significant event in the early stages of disease. BAL levels of HNL are reported to correlate with decreased DLCO and high resolution CT emphysematous scores in "healthy" smokers with subclinical disease [42]. Elevated levels of HNL are also detected in induced sputum from COPD patients [11] suggesting that this method of sampling may also be well suited to evaluation of neutrophil activation at early stages of disease.

Increased numbers of alveolar macrophages (AMs) are reported in BAL samples from smokers *versus* non-smokers [43]. Whether this reflects an early event in the inflammatory process leading to disease is controversial, however. For example, Abboud et al. [44] report that smokers with mild emphysema as evidenced by CT scan have increased numbers of BAL alveolar macrophages compared to smokers without CT evidence of disease while Linden et al. [2] report no difference in BAL cell numbers or composition between smokers with and without airway disease. These findings may reflect differing involvement of AMs in the airway obstruction and alveolar destruction components of the disease. When BAL aliquots were analysed separately, a higher proportion of macrophages was recovered in the alveolar than the bronchial washings [38]. In this context it is of interest that Abboud et al. [44] also observed a relationship between macrophage plasminogen activator and decreased diffusion capacity but not with FEV_1 or FEV_1/FVC ratio. If AMs are indeed more involved in the emphysematous component of the disease then it is unlikely that induced sputum, which exclusively samples the upper airways, will be of value in elucidating their role in the disease process.

Studies on lung tissue have implicated the CD8+ve T lymphocyte as an effector cell in the progression of COPD [41]. Although few BAL studies have specifically focused on the lymphocyte component, where subset analysis has been carried out increased BAL CD8+ve lymphocytes are reported to be associated with smoking *per se* [45] with no difference being observed between smokers with and without disease [35]. As availability of lung tissue from non-smokers and non-diseased smokers is extremely limited and yield of lymphocytes by induced sputum is very low, further BAL studies are best suited to clarify the involvement of CD8+ve T lymphocytes in the disease process.

As outlined earlier, the role of eosinophils in COPD is controversial and may reflect an asthmatic component in the airways. Although the yield of eosinophils in BAL is low (usually < 1% of inflammatory cells), an increased proportion of these granulocytes is observed in BAL from patients compared to healthy controls [46] and ex-smokers [36]. Increased BAL levels of the eosinophil degranulation product ECP are also observed [36] and this marker is reported to be significantly correlated with reduced FEV_1 in patients with airway obstruction [37]. As the yield of eosinophils from induced sputum is greater than from BAL [36], use of this technique is probably best suited to further studies on the contribution of this cell to airway inflammation in COPD.

BAL evaluation of soluble inflammatory mediators in COPD airways

Arising from the association of neutrophils with disease progression in COPD, several studies have evaluated levels of neutrophil chemoattractants in BAL. Raised levels of GM-CSF [47] and IL-8 [40, 48] are seen in patients with obstructive disease. Increased IL-8 is also seen in smokers, being particularly elevated in the bronchial component of the lavage where levels are correlated with neutrophil count [42, 49]. A similar correlation between BAL IL-8 levels and neutrophils is observed in COPD patients where levels are inversely correlated with airflow obstruction [40]. In a recent study by Tanino et al. [48], who evaluated BAL for levels of three neutrophil chemotaxins (epithelial neutrophil activating protein 78 (ENA-78), LTB_4, and IL-8), only IL-8 levels were found to distinguish between current smokers with and without emphysema. These studies strongly suggest that IL-8 is the major neutrophil chemoattractant in the COPD airways and that it is particularly concentrated in the upper airways. In this context it is of interest that Tanino et al. found no difference in IL-8 expression or production in alveolar macrophages from smokers with or without emphysema [48] while Mio et al. [49] have demonstrated that bronchial epithelial cells release IL-8 in response to cigarette smoke. Thus, for further evaluation of neutrophil chemotaxis and related inflammatory mediators, induced sputum may be the most appropriate sampling method.

A range of other inflammatory mediators, including interleukin (IL)-1β, IL-6, TNF-α, monocyte chemoattractant protein 1 (MCP-1), GRO-α and macrophage inflammatory protein-1α (MIP-1α) have also been assessed in BAL from smokers and patients with obstructive disease [20, 40, 48]. None have demonstrated a relationship with disease similar to that seen with IL-8.

BAL evaluation of proteases and antiproteases in the COPD airways

In light of the elastase-anti-elastase hypothesis on the pathogenesis of emphysema, early BAL studies focused on the elastase-anti-elastase balance in BAL fluid from COPD patients and smokers. These studies, however, failed to give consistent findings. Carp et al. [50] reported decreased elastase inhibitory activity and decreased levels of α_1-proteinase inhibitor in BAL fluid from smokers. By comparison, Smith et al. [51] found no differences in antiprotease levels or functional inhibitory capacities between BALs from smokers and controls. Similarly, Morrison et al. [39] found no difference in the capacity of BAL fluids from patients with chronic bronchitis or alpha 1-proteinase inhibitor deficiency to inhibit neutrophil elastase. Reported observations on elastolytic activity were more consistent, with increased activity being observed in smokers with and without obstructive disease [51, 52]. Later studies demonstrated an increase in neutrophil elastase-α_1-proteinase inhibitor complex levels in BAL from smokers with subclinical emphysema and found that this was correlated with accelerated decline in FEV_1 on follow-up [53].

BAL levels of matrix metalloproteinases have also been evaluated, with increases in collagenases (MMPs 1and 8) and gelatinase B (MMP-9) being observed [25, 54]. Comparison with elastase levels suggest that BAL collagenase activity may be a better indicator of presence of disease than elastase [25]. To date, however, studies on the balance between BAL levels of MMPs and their inhibitors, the TIMPs, have not been reported.

Conclusion

The studies reviewed indicate that BAL and induced sputum, by sampling different compartments in the airways, give complementary information on the inflammatory process in COPD. The observed associations between degree of airway obstruction, neutrophil numbers and IL-8 levels in induced sputum and the bronchial component of BAL point to induced sputum as the sampling method of choice for evaluating the neutrophilic component of the inflammatory process in COPD. This method is eminently suitable for studies designed to tease out the crucial differences between smokers who develop disease and those who do not, as it can be safely and repeatedly employed to sample healthy smokers over prolonged periods of time.

The safety and level of ease with which repeat induced sputum samples can be obtained, even from patients with severe disease, also make this procedure the most promising for evaluation of airway disease progression and the effect of new therapies.

A significant drawback in the induced sputum procedure, however, is the use of DTT to solubilise mucus plugs. This can significantly effect measurement of soluble mediators and is likely to influence the behaviour of recovered cells. Where cell function or soluble mediators are to be assessed in induced sputum samples, preliminary studies to determine the effect of DTT on the experimental and analytical methods employed are absolutely essential.

While induced sputum is the sampling technique of choice for evaluating the airway obstruction component of the disease, it can shed little light on the emphysematous process. BAL remains the method of choice for studies aimed at determining the early pathological events involved in alveolar destruction.

References

1   Pozzi E, De Rose V, Rennard SI, Fabbri LM (1990) Clinical guidelines and indications

for bronchoalveolar lavage (BAL): chronic bronchitis and emphysema. *Eur Respir J* 3: 961–969

2   Linden M, Rasmussen JB, Piitulainen E, Larsson M, Brattsand R (1990) Inflammatory indices for chronic bronchitis and chronic obstructive airway disease. Cell populations in bronchial and bronchoalveolar lavage. *Agents Actions* (Suppl) 30: 183–197

3   Fahy JV, Boushey HA, Lazarus SC, Mauger EA, Cherniac RM, Chinchilli VM, Craig TJ, Drazen JM, Ford JG, Fish JE et al (2001) Safety and reproducibility of sputum induction in asthmatic subjects in a multicenter study. *Am J Respir Crit Care Med* 163: 1470–1475

4   Bhowmik A, Seemungal TAR, Sapsford RJ, Devalia JL, Wedzicha JA (1998). Comparison of spontaneous and induced sputum for investigation of airway inflammation in chronic obstructive pulmonary disease. *Thorax* 53: 953–956

5   Leigh TR, Parsons P, Hume C, Husain OAN, Gazzard B, Collins JV (1989) Sputum induction for diagnosis of Pneumocystis Carinii pneumonia. *Lancet* 2(8656): 205–206

6   Pin I, Gibson PG, Kolendowicz R, Girgis-Gabardo A, Denburg JA, Hargreave FE, Dolowich J (1991) Use of induced sputum cell counts to investigate airway inflammation in asthma. *Thorax.* 47: 25–29

7   Vlachos-Mayer H, Leigh R, Sharon RF, Hussack P, Hargreave FE (2000) Success and safety of sputum induction in the clinical setting. *Eur Respir J* 16: 997–1000

8   Hunter CJ, Ward R, Woltmann G, Wardlaw AJ, Pavord ID (1999) The safety and success rate of sputum induction using a low output ultrasonic nebuliser. *Respir Med* 93: 345–348

9   Gershman NH, Wong HH, Liu JT, Mahlmeister MJ, Fahy JV (1996) Comparison of two methods of collecting induced sputum in asthmatic subjects. *Eur Respir J* 9: 2448–2453

10 Paggiaro PL, Chanez P, Holz O, Ind PW, Djukanovicz R, Maestrelli P, Sterk PJ (2002) Sputum induction *Eur Respir J* 20 (Suppl 37): 3s–8s

11 Keatings VM, Barnes PJ (1997) Granulocyte activation markers in induced sputum: comparison between chronic obstructive pulmonary disease, asthma, and normal subjects. *Am J Respir Crit Care Med* 155: 449–453

12 Fahy JV, Liu J, Wong H, Boushey HA (1993) Cellular and biochemical analysis of induced sputum from asthmatic and from healthy subjects. *Am Rev Respir Dis* 147: 1126–1131

13 Efthidimiadis A, Spanavello A, Hamid Q, Kelly MM, Linden M, Louis R, Pizzichini MMM, Pizzichini E, Ronchi C, Van Overveld F, Djukanovic R (2002) Methods of sputum processing for cell counts, immunocytochemistry and *in situ* hybridization. *Eur Respir J* 20 (Suppl 37): 19s–23s

14 Keatings VM, Collins PD, Scott DM, Barnes PJ (1996) Differences in interleukin-8 and tumour necrosis factor-α in induced sputum from patients with chronic obstructive pulmonary disease and asthma. *Am J Resp Crit Care Med* 153: 530–534

15 Gibson PG, Woolley KL, Carty K, Murree-Allen K, Saltos N (1998) Induced sputum ECP measurement in asthma and COPD. *Clin Exp Allergy* 28: 1081–1088

16 Louis RE, Cataldo D, Buckley MG, Sele J, Henket M, Lau LC, Bartsch P, Walls AF,

Djukanovicz R (2001) Evidence of mast cell activation in a sub-set of patients with eosinophilic COPD. *Eur Respir J* 20: 325–331

17 Stockley RA, Bayley DL (2000) Validation of assays for inflammatory mediators in sputum. *Eur Respir J* 15: 778–781

18 Kelly MM, Keatings V, Leigh R, Peterson C, Shute J, Venge P, Djukanovicz R (2002) Analysis of fluid phase mediators. *Eur Respir J* 20 (Suppl 37): 24s–39s

19 Hill A, Bayley D, Stockley RA (1999) The interrelationship of sputum inflammatory markers in patients with chronic bronchitis. *Am J Respir Crit Care Med* 160: 893–898

20 Traves SL, Culpitt SV, Russell REK, Barnes PJ, Donnelly LE (2002) Increased levels of the chemokines GRO-α and MCP-1 in sputum from patients with COPD. *Thorax* 57: 590–595

21 Morrison D, Streiter RM, Donnelly SC, Burdick MD, Kunkel SL, Mac Nee W (1998) Neutrophil chemokines in BAL fluid and leukocyte conditioned medium from non-smokers and smokers. *Eur Respir J* 12: 1067–1072

22 Thomas PS, Yates DH, Barnes PJ (1995) Tumor necrosis factor-alpha increases airway responsiveness and sputum neutrophilia in normal human subjects. *Am J Respir Crit Care Med* 152: 76–80

23 Armstrong L, Jordan N, Millar A (1996) Interleukin 10 (IL-10) regulation of tumour necrosis factor alpha (TNF-alpha) from human alveolar macrophages and peripheral blood monocytes. *Thorax* 51: 143–149

24 Takanashi S, Hasegawa Y, Kanehira Y, Yamamoto K, Fujimoto K, Satoh K, Okamura K (1999) Interleukin-10 level in sputum is reduced in bronchial asthma, COPD and in smokers. *Eur Respir J* 14: 309–314

25 Finlay GA, Russell KJ, McMahon KJ, D'Arcy EM, Masterson JB, FitzGerald MX, O'Connor CM (1997) Elevated levels of matrix metalloproteinases in bronchoalveolar lavage fluid of emphysematous patients. *Thorax* 52: 502–506

26 Vignola AM, Riccobono L, Mirabella A, Profita M, Chanez P, Bellia V, Mautino G, D'Accardi P, Bousquet J, Bonsignore G (1998) Sputum metalloproteinase-9/tissue inhibitor of metalloproteinase-1 ratio correlates with airflow obstruction in asthma and chronic bronchitis. *Am J Respir Crit Care Med* 158: 1945–1950

27 Vignola AM, Bonanno A, Mirabella A, Riccobono L, Mirabella F, Profita M, Bellia V, Bousquet J, Bonsignore G (1998) Increased levels of elastase and alpha1-antitrypsin in sputum of asthmatic patients. *Am J Respir Crit Care Med* 157: 505–511

28 Culpitt SV, Maxiak W, Loukidis S, Nightingale JA, Matthews JL, Barnes PJ (1999) Effect of high dose inhaled steroid on cells, cytokines and proteases in induced sputum in chronic obstructive disease. *Am J Respir Crit Care Med* 160: 1635–1639

29 Cataldo D, Munaut C, Noel A, Frankenne F, BartschP, Foidart JM, Louis R (2000) MMP-2 and MMP-9 linked gelatinolytic activity in the sputum from patients with asthma and chronic obstructive pulmonary disease. *Int Arch Allergy Immunol* 123: 259–267

30 Bhowmik A, Seemungal TA, Sapsford RJ, Wedzicha JA (2000) Relationship of sputum

inflammatory markers to symptoms and lung function changes in COPD exacerbations. *Thorax* 55: 114–120

31 Gompertz S, Bayley DL, Hill SL, Stockley RA (2001) Relationship between airway inflammation and the frequency of exacerbations in patients with smoking related COPD. *Thorax* 56: 36–41

32 Keatings VM, Jatakanon A, Worsdell YM, Barnes PJ (1997) Effects of inhaled and oral glucocorticoids on inflammatory indices in asthma and COPD. *Am J Resp Crit Care Med* 155: 542–548

33 Pizzichini E, Pizzichini MM, Gibson P, Parameswaran K, Gleich G, Berman L, Dolovich J, Hargreave FE (1998) Sputum eosinophilia predicts benefit from prednisolone in smokers with chronic obstructive bronchitis. *Am J Respir Crit Care Med* 158: 1511–1517

34 Brightling R, Monteiro W, Ward R, Parker D, Morgan MDL, Wardlaw AJ, Pavord ID (2000) Sputum eosinophilia and short-term response to prednisolone in chronic obstructive pulmonary disease: a randomized controlled trial. *Lancet* 356: 1480–1485

35 Costabel U, Maier K, Teschler H, Wang YM (1992) Local immune components in chronic obstructive pulmonary disease. *Respiration* 59: 17–19

36 Rutgers SR, Timens W, Kaufmann HF, van der Mark TW, Koeter GH, Postma DS (2000) Comparison of induced sputum with bronchial wash, bronchoalveolar lavage and bronchial biopsies in COPD. *Eur Respir J* 15: 109–115

37 Linden M, Rasmussen JB, Piitulainen E, Tunek A, Larson M, Tegner H, Venge P, Laitinen LA, Brattsand R (1993) Airway inflammation in smokers with nonobstructive and obstructive chronic bronchitis. *Am Rev Respir Dis* 148: 1226–1232

38 Martin TR, Raghu G, Maunder RJ, Springmeyer SC (1985) The effects of chronic bronchitis and chronic air-flow obstruction on lung cell populations recovered by bronchoalveolar lavage. *Am Rev Respir Dis* 132: 254–260

39 Morrison HM, Kramps JA, Burnett D, Stockley RA (1987) Lung lavage fluid from patients with alpha 1-proteinase inhibitor deficiency or chronic obstructive brochitis: anti-elastase function and cell profile. *Clin Sci* 72: 373–381

40 Soler N, Ewig S, Torrs A, Filella X, Gonzalez J, Zaubet A (1999) Airway inflammation and bronchial microbial patterns in patients with stable chronic obstructive pulmonary disease. *Eur Respir J* 14: 1015–1022

41 Maestrelli P, Saetta M, Mapp CE, Fabbri LM (2001) Remodeling in response to infection and injury. Airway inflammation and hypersecretion of mucus in smoking subjects with chronic obstructive pulmonary disease. *Am J Respir Crit Care Med* 164: S76–S80

42 Ekberg-Jansson A, Andersson B, Bake B, Boijsen M, Enanden I, Rosengren A, Skoogh BE, Tylen U, Venge P, Lofdahl CG (2001). Neutrophil-associated activation markers in healthy smokers relates to a fall in DL(CO) and to emphysematous changes on high resolution CT. *Respir Med* 95: 363–373

43 Pratt S, Finley T, Smith M, Ladman A (1969) A comparison of alveolar macrophages and pulmonary surfactant obtained from the lungs of human smokers and non-smokers by endobroncchial lavage. *Anat Rec* 163: 497–506

44 Abboud RT, Ofulue AF, Sansores RH, Muller NL (1998) Relationship of alveolar macrophage plasminogen activator and elastase activities to lung function and CT evidence of emphysema. *Chest* 113: 1257–1263

45 Ekberg-Jansson A, Andersson B, Avra E, Nilsson O, Lofdahl CG (2000) The expression of lymphocyte surface antigens in bronchial biopsies, bronchoalveolar lavage cells and blood cells in healthy smoking and never-smoking men, 60 years old. *Respir Med* 94: 264–272

46 Balbi B, Aufiero A, Pesci A, Oddera S, Zanon P, Rossi GA, Olivieri D (1994) Lower respiratory tract inflammation in chronic bronchitis. Evaluation by bronchoalveolar lavage and changes associated with treatment with Immucytal, a biological response modifier. *Chest* 106: 819–826

47 Balbi B, Bason C, Balleari E, Fiasella F, Pesci A, Ghio R, Fabiano F (1997) Increased bronchoalveolar granulocytes and granulocyte/macrophage colony-stimulating factor during exacerbations of chronic bronchitis. *Eur Respir J* 10: 846–850

48 Tanino M, Betsuyaku T, Takeyabu K, Tanino Y, Yamaguchi E, Miyamoto K, Nishimura M (2002) Increased levels of interleukin-8 in BAL fluid from smokers susceptible to pulmonary emphysema. *Thorax* 57: 405–411

49 Mio T, Romberger DJ, Thompson AB, Robbins RA, Heires A, Rennard SI (1997) Cigarette smoke induces interleukin-8 release from human bronchial epithelial cells. *Am J Respir Crit Care Med* 155: 1770–1776

50 Carp H, Miller F, Hoidal JH, Janoff A (1982) Potential mechanism of emphysema: alpha 1-proteinase inhibitor recovered from lungs fo cigarette smokers contains oxidized methionine and has decreased elastase inhibitory capacity. *Proc Natl Acad Sci USA* 79: 2041–2045

51 Smith SF, Guz A, Cooke NT, Burton GH, Tetley TD (1985) Extracellular elastolytic activity in human lung lavage: a comparative study between smokers and non-smokers. *Clin Sci* 69: 17–27

52 Terpstra GK, De Weger RA, Wassink GA, Kreuknit J, Huidekoper HJ (1987) Changes in alveolar macrophage enzyme content and activity in smokers and patients with chronic obstructive lung disease. *Int J Clin Pharmacol Res* 7: 273–277

53 Betsuyaku T, Nishimura M, Takeyabu K, Tanino M, Miyamoto K, Kawakami Y (2000) Decline in FEV(1) in community-based older volunteers with higher levels of neutrophil elastase in bronchoalveolar lavage fluid. *Respiration* 67: 261–267

54 Segura-Valdez L, Pardo A, Gaxiola M, Uhal BD, Becerril C, Selman M (2000) Up-regulation of gelatinases A and B, collagenases 1 and 2, and increased parenchymal cell death in COPD. *Chest* 117: 684–694

Exhaled breath markers in COPD

Sergei A. Kharitonov and Peter J. Barnes

Department of Thoracic Medicine, National Heart and Lung Institute, Imperial College London, Royal Brompton Hospital, Dovehouse Street, London SW3 6LY, UK

Non-invasive biomarkers of inflammation in asthma and COPD

The need to monitor inflammation in the lungs has led to the exploration of exhaled gases and condensates, that may assist in differential diagnosis of pulmonary diseases, assessment of disease severity and response to treatment [1–3]. A proof-of-concept study often will involve an examination of effects on a surrogate marker that is not guaranteed to produce a quick readout, such as effects on, for example neutrophilia or neutrophil activation in chronic obstructive pulmonary disease (COPD).

In addition to the practical aspects of the examination of surrogate markers in the assessment of drug efficacy in the treatment of asthma or COPD patients, it is not clear how the benefits of anti-inflammatory treatment for example, will be manifested in patients with these conditions and, therefore, what indexes should be examined clinically.

Exhaled breath

Nitric oxide (NO)

NO is the most extensively studied exhaled marker and abnormalities in exhaled and nasal NO have been documented in several lung diseases, particularly asthma [4]. Exhaled NO and nitrite/nitrate levels in breath condensate can be used to monitor dose-dependent onset and duration of action of corticosteroids [5], and are valuable parameters to monitor complex NO biochemistry in the clinic (see Tab. 1).

An interesting method of measuring exhaled NO at several exhalation flow rates has recently been described [6]. Peripheral airways/alveolar region may be the predominant source of elevated exhaled NO in COPD [7, 8], while an increased exhaled NO level in asthma is mainly derived from the larger airways [9, 10]. This novel technique can be used not only to monitor the disease, but the effect of treatment with NO modulators, which may have different mechanisms and sites of their actions.

Table 1 - Exhaled gases, breath temperature and bronchial blood flow in COPD

	Control (non-smokers)[0]	Control (smokers)[1]	Mild[2]	Disease Moderate[3]	Severe[4]	Effect of disease	Effect of steroids	Refs.
COPD								
NO								
	7	2.9 (0.5)		6.3 (0.6) ex-smokers 4.2 (0.4) current smokers 8.2 (1.2) IS 5 (0.4) no IS	12.1 (1.5) unstable	117% ↑ 1-3 ex-smokers 45% ↑ 1-current smokers 317% ↑ 1-4	38% ↓ 3 IS vs. no IS	[43]
	6.5 (0.6)	3.3 (0.4)		12 (1.0) ex-smokers 7.6 (1.1) smokers		264% ↑ 1-3 ex-smokers 139% ↑ 1-3 ex-smokers	No effect	[44]
	9.4 (0.8)	4.6 (0.4)		25.7 (3.0) ex-smokers 10.2 (1.4) current smokers 28.3–22.7 FP 1000 μg 28.3–31.0 placebo		459% ↑ 1-3 ex-smokers 122% ↑ smoker 19% ↓ FP 1000 μg No effect placebo		[45] [46]
				13.7 (1.9)–12.9 (2.9) BDP 800 μg			No effect	[7]
				12.5 (3.6)–13.3 (3.8) placebo			No effect	
Multiple exhalation flow technique (larger airway NO, JNO, pL/s or nL/s)								
	667 (24)	477 (23)		617 (61)	978 (74)	28% ↓ 0-1 7% ↓ 0-3 46% ↑ 0-4		[47]

Table 1 (continued)

	Control (non-smokers)[0]	Control (smokers)[1]	Mild[2]	Disease Moderate[3]	Severe[4]	Effect of disease	Effect of steroids	Refs.
Multiple exhalation flow technique (small airway NO, Calv, ppb)								
	1.5 (0.07)	2.45 (0.3) At risk (GOLD)		3.7 (0.1)	4.76 (0.2)	63% ↑ 0–1 51% ↑ 1–3 94% ↑ 1–4		[47]
	2 (1)			4 (2)		100% ↑ 0–3		[48]
Ethane	0.88 (0.09)			2.77 (0.25) no ICS 0.48 (0.05) ICS		215% ↑ 0–3 no ICS	83% ↓ 3 no ICS-ICS	[12]
Exhaled breath temperature (Δ°C/s)								
	4.0 (0.26)			1.86 (0.15)		54% ↓ 0–3	No effect	[25]

8-iso, 8-isoprostane; cys-LT, cysteinyl leukotrienes (LTC$_4$/D$_4$/E$_4$); LTB$_4$, leukotriene B$_4$; PGE$_2$, prostaglandin E$_2$; LTE$_4$, leukotriene E$_4$; PGD$_2$, prostaglandin D$_2$; PGF$_{2\alpha}$ prostaglandin F$_{2\alpha}$; TXB$_2$, tromboxane B$_2$; MDA, malondialdehyde; NT, nitrotyrosine; an effect of disease or steroids is expressed as the comparison between, for example, control ([0]) and mild COPD ([1]), [0–1]; Bud, budesonide; BDP, beclomethasone dipropionate; FP, fluticasone; data are mean ± (SEM), or [SD], or (95% CI); Es, EcoScreen condenser; RT, Rtube condenser; G, glass condenser; CA, colorimetric assay; SPh, spetrophotometry; HPLC, high performance liquid chromatography; MS, masspectrometer.

Hydrocarbons
Exhaled ethane is increased in normal smokers and COPD patients [11]. Increased levels of volatile organic compounds in exhaled breath could be used as biochemical markers of exposure to cigarette smoke and oxidative damage caused by smoking. In fact, there is a correlation between the ethane levels and the degree of airway obstruction in COPD [12].

Exhaled breath condensate (EBC)

EBC is collected by cooling or freezing exhaled air and is totally non-invasive. Although the collection procedure has not been standardized, there is strong evidence that abnormalities in condensate composition may reflect biochemical changes of airway lining fluid [4]. Potentially, EBC can be used to measure the targets of modern therapy in clinical trials and monitor asthma and COPD in clinic (Tab. 2).

Hydrogen peroxide
Activation of inflammatory cells, including neutrophils, macrophages and eosinophils, result in an increased production of O_2^- and formation H_2O_2. As H_2O_2 is soluble, increased H_2O_2 in the airway equilibrates with air and can be detected in exhaled breath condensate. Thus exhaled H_2O_2 has potential as a marker of oxidative stress in the lungs.

Cigarette smoking causes an influx of neutrophils and other inflammatory cells into the lower airways and five-fold higher levels of H_2O_2 have been found in exhaled breath condensate of smokers than in non-smokers. Levels of exhaled H_2O_2 are increased compared to normal subjects in patients with stable COPD and are further increased during exacerbations.

Tyrosine, nitrotyrosine, nitrite, nitrate, reactive nitrogen species
A significant proportion of NO is consumed by chemical reactions in the lung leading to formation of nitrite, nitrate, and S-nitrosothiol in the lung epithelial lining fluid. Elevated levels of S-nitrosothiols in exhaled breath condensate has been demonstrated in COPD [13].

Eicosanoids (prostanoids, leukotrienes, 8-isoprostanes)
Eicosanoids are potent mediators of inflammation responsible for vasodilatation/vasoconstriction, plasma exudation, mucus secretion, bronchoconstriction/bronchodilatation, cough and inflammatory cell recruitment. Exhaled prostanoids, for

example PGE$_2$ and PGF$_{2\alpha}$, are detectable in exhaled breath condensate and markedly increased in patients with COPD, whereas these prostaglandins are not significantly elevated in asthma [14, 15]. In contrast, TxB$_2$ is increased in asthma but not detectable in normal subjects of in patients with COPD [16].

Detectable levels of LTB$_4$, C$_4$, D$_4$, E$_4$ and F$_4$ have been reported in exhaled condensate of asthmatic and normal subjects [17, 18]. The levels of leukotrienes LTE$_4$, C$_4$, D$_4$ in breath condensate are elevated significantly in patients with moderate and severe asthma [17], and steroid withdrawal in moderate asthma leads to worsening of asthma and further increase in exhaled NO and the concentration of LTB$_4$, LTE$_4$, LTC$_4$, LTD$_4$ in exhaled condensate [19]. LTB$_4$ concentrations are increased in exhaled breath condensate of patients with stable COPD [16], COPD exacerbations [20] as well as in moderate and severe asthma [17]. This suggests that LTB$_4$ may also be involved in exacerbations of asthma and may contribute towards neutrophil recruitment.

Isoprostanes are prostanoids formed by free radical-catalyzed lipid peroxidation of arachidonic acid. They are not simply markers of lipid peroxidation but also possess biological activity, and could be mediators of the cellular effects of oxidant stress and a reflection of complex interactions between the RNS and ROS. 8-isoprostane levels are approximately doubled in mild asthma compared with normal subjects, and increased by about three-fold in those with severe asthma, irrespective of their treatment with corticosteroids. The relationship to asthma severity is a useful aspect of this marker, in contrast to exhaled NO. The relative lack of effect of corticosteroids on exhaled 8-isoprostane has been confirmed in a placebo-controlled study with the two different doses of inhaled steroids [21]. This provides evidence that inhaled corticosteroids may not be very effective in reducing oxidative stress. Exhaled isoprostanes may be a better means of reflecting disease activity than exhaled NO. The concentration of 8-isoprostane in exhaled condensate is also increased in normal cigarette smokers, but to a much greater extent in COPD patients [22]. Interestingly, exhaled 8-isoprostane is increased to a similar extent in COPD patients who are ex-smokers as in smoking COPD patients, indicating that the exhaled isoprostanes in COPD are largely derived from oxidative stress from airway inflammation, rather than from cigarette smoking.

Exhaled breath temperature and bronchial blood flow

Exhaled breath temperature and bronchial blood flow are other quantitative markers of airway inflammation, as vascular hyperperfusion plays an important role in tissue inflammation and airway remodelling. We have developed and validated a novel technique for the measurement of exhaled breath temperature and bronchial blood flow [23] in asthma [24] and COPD [25]. Exhaled breath temperature and blood flow is increased in asthma to inflammatory new vessel formation and vasodi-

Table 2 - Biomarkers of exhaled breath condensate in COPD

Biomarker	Control (non-smokers)[0]	Control (smokers)[1]	Mild[2] COPD	Moderate[3] COPD	Severe[4]	Effect of disease/smoking	Effect of steroids	Collection/analysis	Refs.
8-iso (pg/ml)	10.8 (0.8)	24.3 (2.6)		39.9 (3.1)[ES] 45.3 (3.6)[CS]		125% ↑ [0-1] 269% ↑ [0-3ES] 319% ↑ [0-3CS]	No effect	Es/ELISA	[22]
	6.2 (0.4)			13.0 (0.9)[E] 9.0 (0.6)[Ant 2 w]		110% ↑ [0-3]		Es/ELISA	[20]
H_2O_2 (µM)	0.029 (0.012)			2.6 (1.9–3.5)			No effect 8% ↓ BDP vs.placebo	CrC/HP/CA	[49]
				0.205 (0.054)[stable] 0.6 (0.075)[exacerbation]		607% ↑ [0-3 stable] 1968% ↑ [0-3 exacerbation]		G/HP/CA	[50]
cys-LT (pg/ml)									
LTE$_4$ (pg/ml)	15.5 (11–27)			23 (9–31)[SN] 19 (9–42)[ST]		No effect	No effect	Es/ELISA	[16]
LTB$_4$ (pg/ml)	6.1 (0.3) 38.1 (31.2–53.6)	9.4 (0.4)		100 (73.5–145)[SN]		54% ↑ [0-1] 163% ↑ [0-3]	No effect	Es/ELISA Es/ELISA	[51] [16]
	7.7 (0.5)			99 (57.9–170)[ST] 15.8 (0.9)[E] 9.9 (0.9)[Ant 2 w]		160% ↑ [0-3] 105% ↑ [0-E3]		Es/ELISA	[20]
PGE$_2$ (pg/ml)	44.3 (30.2–52.1)			98 (57–128)[SN] 93 (52–157)[ST]		122% ↑ [0-3SN] 111% ↑ [0-ST]	No effect	Es/ELISA	[16]

Table 2 (continued)

Biomarker	Control (non-smokers)[0]	Control (smokers)[1] Mild[2]	COPD Moderate[3]	Severe[4]	Effect of disease/smoking	Effect of steroids	Collection/analysis	Refs.
PGD_2 (pg/ml)	10.0 (8–15) detectable in 40% samples		11.2 (8–15) SN detectable in 30% samples 11.4 (8–14) ST detectable in 28% samples		No effect	No effect	Es/ELISA	[16]
$PGF_{2\alpha}$ (pg/ml)	8.9 (5.9–10.9) detectable in 33% samples		15 (10–19) SN detectable in 25% samples 14 (10–21) ST detectable in 24% samples		68% ↑ [0–3] 57% ↑ [0–3]	No effect	Es/ELISA	[16]
TXB_2 (pg/ml)	Undetectable						Es/ELISA	
IL-6 (pg/ml)	2.6 (0.2)	5.6 (1.4)			115% ↑ [0–1]		Es/ELISA	[51]
Adenosine (nM)								
SNO (μM)	0.11 (0.02)	0.46 (0.09)	0.24 (0.04) 43% on ICS		318% ↑ [0–1] 118% ↑ [0–3]	Possible effect (↓)	G/CA/SPh	[13]
NT (ng/ml)								
MDA (nmol/l)	12.1 (1.8)	35.6 (4.0)	57.2 (2.4)		194% ↑ [0–1] 372% ↑ [0–3] (35% current smokers)	No effect	G/HPLC-MS	[52]

Abbreviations see Table 1.

latation, but not in COPD, which may reflect the changes of bronchial blood flow and tissue remodelling.

Standardisation and reproducibility of biomarkers collection and measurements

Airway inflammation in asthma has been shown to exhibit considerable biologic variability yet it is not measured directly and routinely in clinical practice. Several approaches are currently used to measure airway inflammation in COPD. They are, however, either invasive or semi-invasive (bronchoscopy and sputum induction), or indirect (PC_{20} and lung function), or may be affected by patient's perception or bronchodilators. The reproducibility of these approaches is also variable.

Exhaled breath
The changes in serial exhaled NO (FE_{NO}), as a loss-of-control-marker [26] have higher predictive values for diagnosing deterioration of asthma than single measurements.

It can be argued, however, that these changes in FE_{NO} may be due to measurement error and/or the natural variability of airway inflammation over time. Therefore, design and interpretation of clinical studies in asthma and the use of FE_{NO} in routine clinical practice depend greatly on reproducibility and safety of FE_{NO} measurements. Reproducibility of FE_{NO} measurements within a single day in both adults (intraclass correlation coefficient (ICC) 0.94) and children (ICC 0.94) is superior to any conventional methods of airway inflammation monitoring in asthma. This adds significantly to other major advantages of FE_{NO} measurements, such as their strong association with airway inflammation [1], even in non-symptomatic asthma patients [27], their high sensitivity to steroid treatment, insensitivity to β_2-agonists and non-invasiveness.

Repeated FE_{NO} measurements, therefore, can be used much more frequently and will not disturb the system, in contrast to the invasive or semi-invasive procedures currently used in clinical research to monitor inflammation status [1]. There are several important practical implications for the FE_{NO} measurements regarding the data comparison of spirometry *versus* FE_{NO} examination. Firstly, we have shown that high reproducibility of FE_{NO} measurements in both children and adults may allow the medical practitioner to perform two instead of three exhalations to obtain the reliable results. This may be of great advantage, as it will shorten the time needed for the measurement procedure. Secondly, because the FE_{NO} measurements by NIOX are fully automated and incorrect exhalation manoeuvres by a patient (shorter than 10 s or above the certain limits of the exhaled flow) will not be accepted by the analyser, the staff training procedure can be minimal. Finally, the advantage of

FE$_{NO}$ measurements is that it does not require an extra encouragement, as it may be in case of PEF measurements.

There is also no "learning effect" or systematic error of serial FE$_{NO}$ measurements. The simplicity and high reproducibility of FE$_{NO}$ measurements in our study are probably the major reasons for this. The mean pooled SD of all measurements was 2.1 ± 1.25 ppb [28]. These results suggest that if a patient's exhaled FE$_{NO}$ levels change more than 4 ppb between sessions, it is more likely due to the inflammatory process rather than inaccuracy of the analyser. This finding is valuable for potential use of FE$_{NO}$ in routine clinical practice. Short-term monitoring, when the measurements of airway inflammation are made more often, for example every day or twice a day as in the case of PEF, is particularly important. This is because of a recent trend towards use of lower doses of inhaled corticosteroids in combination with long-acting β$_2$-agonist (LABA) when anti-inflammatory and clinical effect of combination treatment may be seen within hours and days. Another example is the use of specific inducible NO synthase (iNOS) inhibitors, which may be a potential additional treatment of severe asthma, COPD or arthritis, when the effect of iNOS inhibitors may be seen within minutes [29, 30] and last for 72 hours [30].

Exhaled condensate
Several methods of condensate collection have been described. The most common approach is to ask the subject to breathe tidally *via* a mouthpiece through a non-rebreathing valve in which inspiratory and expiratory air is separated. During expiration the exhaled air flows through a condenser, which is cooled to 0 °C by melting ice [31], or to –20 °C by a refrigerated circuit [32], and breath condensate is then collected into a cooled collection vessel. A low temperature may be important for preserving labile markers as lipid mediators during the collection period, which usually takes between 10–15 minutes to obtain 1–3 ml of condensate. Exhaled condensate may be stored at –70 °C and is subsequently analyzed by gas chromatography and/or extraction spectrophotometry, or by immunoassays (ELISA). Exhaled breath condensate collection is a simple and well tolerated method that can be safely used in both children [18, 33] and adults [31, 32].

The presence of high concentrations of nitrite/nitrate from the diet may potentially affect NO-related markers in condensate. It is therefore important to minimize and monitor salivary contamination. Subjects should rinse their mouth before collection and to keep the mouth dry by periodically swallowing their saliva.

Sample-size determination

Sample-size determination is often an important step in planning such studies. According to our data (Fig. 1), a small number (between seven and 20) of asthmat-

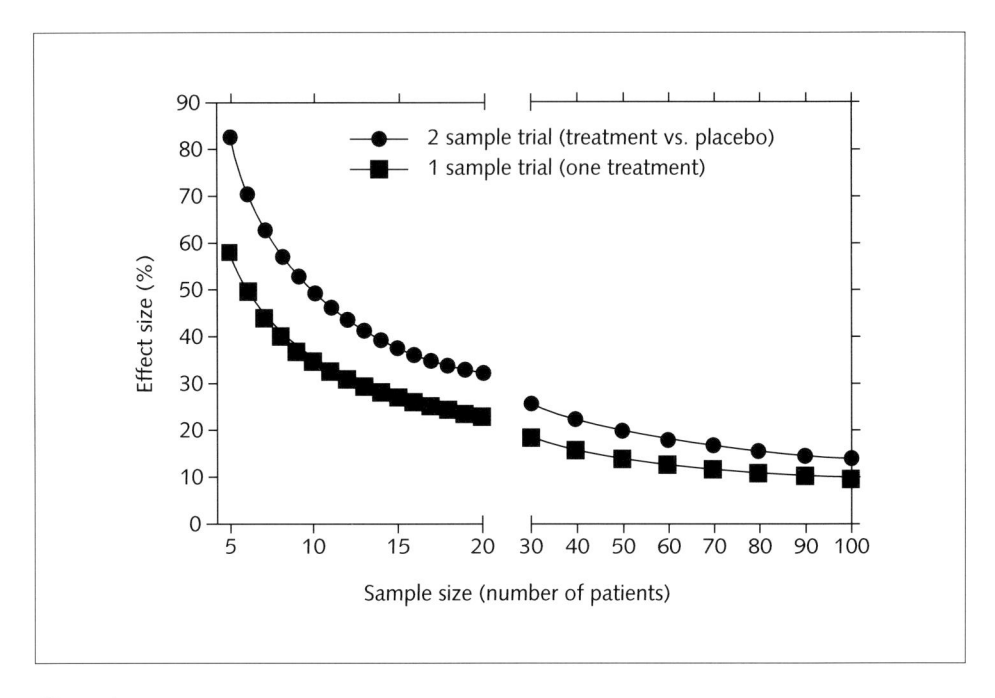

Figure 1
The sample size estimates are based on the variability in the outcomes over the two sample (treatment versus *placebo) and one sample (one treatment) study [34].*

ic subjects, either adults or children, will be required to demonstrate 25–80% effect of a studied drug in a clinical trial. Based on the knowledge of the individual variability of FE_{NO} measurements, like individual peak expiratory flows, individual FE_{NO} values should be established and monitored, and when the levels are above or below a certain reference level, steroid treatment should be either reduced or increased [34].

Corticosteroids and biomarkers in COPD

Determining the appropriate dose of inhaled steroids only by reference to symptoms and lung function, both of which are distant and non-specific markers of the underlying inflammatory process, may be over simplistic. Lung function, a measure of airway calibre, provides little information on airway hyper responsiveness and degree of airway inflammation in asthma and is mostly irreversible in COPD. It is a fact that despite normal or near-normal lung function, persisting inflammation of the airway often exists, and significant reductions in basement membrane thickness did not occur until after three months of treatment with high-dose inhaled corticosteroids (ICS), long before which maximal improvement in lung function would have been achieved.

There is the need for early and long-term intervention studies with ICS alone and ICS + LABA in COPD against more direct indices of inflammation.

Effect of corticosteroids

Acute effect (onset of action)

There have been no direct measurements of acute ICS effects on airway inflammation and microvascular permeability in asthma and COPD, although the newer ICS are more airway selective, have greater receptor affinity (FP), or may have faster onset of action owing to the smaller particle size, for example hydrofluoroalkane beclomethasone. A rapid, topical anti-inflammatory action may require rapid means of monitoring its anti-inflammatory effect, thus minimizing the risk of systemic effects.

Exhaled NO behaves as a "rapid response" marker, which is extremely sensitive to steroid treatment, as it may be significantly reduced even after six hours following a single treatment with a nebulized budesonide [35], or within 2–3 days [5, 36] after regular treatment with inhaled corticosteroids. The onset of action of inhaled budesonide on exhaled NO was dose-dependent, both within the initial phase (first 3–5 days of treatment) and during treatment weeks 1, 2, and 3 [5]. The reduction in exhaled NO during the first 3–5 days was thus significantly faster in the group receiving 400 μg/day budesonide than in the group receiving 100 μg/day. The mean difference between the effect of 100 and 400 μg budesonide was –1.55 ppb/day. There was no effect of either treatment or placebo on exhaled levels of CO, either during the onset or cessation of their action [5].

The above effects were evident without concomitant improvements in lung function. This disconnection between lung function and inflammatory markers may reflect the short-term positive effects of ICS, which may occur within even 3–5 days [5], as improvements in lung function may take longer to occur. Furthermore, it remains unclear whether this "rapid improvement" in inflammatory markers would translate into a commensurate improvement in airway remodelling or exacerbations over the longer term.

The reduction in nitrite/nitrate levels in exhaled condensate during the onset and cessation of action of ICS may be also fast but not dose-dependent [5]. This may be a reflection of a rather complex process of nitrite/nitrate formation and metabolism, and the greater variability in the measurements compared with exhaled NO.

Exhaled CO has been suggested as a marker to monitor airway inflammation in asthma [37]. However, we did not find any changes in either exhaled NO or CO levels three and six hours after a single dose of 100 or 400 μg budesonide [5].

An acute vasoconstriction effect of ICS can be monitored by bronchial blood flow [24, 38], as demonstrated by the skin blanching test. In fact, it may be called "the lung blanching test", perhaps. A single dose of inhaled budesonide (800 μg) caused a rapid (within 30 min), significant but transient (recovery at 60 min) reduction in bronchial blood flow (Paredi et al., unpublished observation) measured by a novel method we have recently developed [23]. Although the maximal FP-induced (880 μg) reduction in airway mucosal blood flow was seen 30 min after drug inhalation in subjects with and without asthma [39], the recovery was slower (90 min). Both in absolute and relative terms, the maximum decrease in bronchial blood flow was greater in asthma than in normal subjects, and this may be explained by its vasoconstrictive effect and/or reduction of airway microvascular permeability seen after a single dose of ICS.

Cessation of action (recovery)
An important question is how fast exhaled NO levels recover when steroid treatment is stopped. We have shown that exhaled NO levels recovered rapidly during the first 3–5 days in all patients treated with ICS, and the full recovery was completed by the end of the week off treatment. Interestingly, the faster recovery of exhaled NO levels in patients treated with 400 μg/day budesonide was independent of the degree of the reduction in exhaled NO by the end of the third week of treatment [5]. The assumption that the effect of 400 μg/day budesonide would last longer if treatment was stopped does not therefore appear to be true.

Long-term effect
Reactive nitrogen species (RNS) have a number of inflammatory actions and the production of these molecules has been reported to be increased in the airways of patients with COPD and severe asthma. Treatment with steroids resulted in a significant reduction in both nitrotyrosine and iNOS immunoreactivity in sputum cells and correlated with the improvement in FEV_1 of COPD patients [7]. It can be speculated that RNS may be involved in the reversible component of inflammation and the long-term progression of COPD and asthma that is suppressed by steroids, or a combination of iNOS inhibitors and ICS [40]. Interestingly, that COPD patients with a partial bronchodilator response to inhaled salbutamol, elevated exhaled NO levels have a different and better response to ICS treatment than do those without reversible airflow limitation [41].

Steroid treatment may also reduce lipid peroxidation in COPD, as the patients receiving steroid treatment had lower levels of exhaled ethane than the untreated patients [12]. Exhaled 8-isoprostanes are increased in asthma and COPD irrespective of treatment with corticosteroids [42]. The lack of effect of corticosteroids on exhaled 8-isoprostane suggests that ICS may not be very effective in reducing oxidative stress. Alternatively, the lack of effect of corticosteroids on inhibiting oxidative stress could be a function of their dose.

It is well established that exhaled NO will be gradually reduced during the first week of regular treatment with inhaled corticosteroids, with maximal reduction at three or four weeks [1]. The reduction in airway mucosal blood flow after a two-week course of intermediate dose of FP (440 µg daily) was modest (11%) but consistent in patients with asthma [38]. It may be speculated that a longer duration of ICS treatment would have resulted in a further decrease of airway mucosal blood flow in the asthma.

Dose-dependent effect

It is difficult to show a dose-dependent effect of inhaled corticosteroids in clinical studies. Dose adjustment in both clinical practice and clinical research is an important issue, in which high reproducibility of FE_{NO} measurements and sensitivity of FE_{NO} to corticosteroids may substantially reduce the cost of medical care and research.

We have shown that the acute (within first 3–5 days of treatment) and chronic (7–21 days) reduction in exhaled NO is dose-dependent in patients with mild asthma treated with low doses of budesonide [5]. Serial exhaled NO measurements, as we recently suggested [1], may therefore be of use to study the onset and duration of action of ICS and patient compliance. It is still unclear whether exhaled NO levels can be used to study the effect of higher doses of ICS. We did not find any further reduction in NO levels in patients with mild asthma treated with 1600 µg/day budesonide for three weeks, although in a separate study a dose-dependent reduction in NO levels was found in patients treated with 100 and 400 µg/day budesonide for three weeks.

Single dose effect

We did not find any changes in either exhaled NO or CO levels three and six hours after a single dose of 100 or 400 µg budesonide [5].

A single dose of inhaled budesonide (800 µg) caused a rapid (within 30 min), significant but transient (recovery at 60 min) reduction in bronchial blood flow (Paredi et al., unpublished observation) measured by a novel method we have recently developed [23]. Although the maximal FP-induced (880 µg) reduction in airway mucosal blood flow was seen 30 min after drug inhalation in subjects with and without asthma [39], the recovery was slower (90 min). Both in absolute and relative terms, the maximum decrease in bronchial blood flow was greater in asthma than in normal subjects, and this may be explained by its vasoconstrictive effect and/or reduction of airway microvascular permeability seen after a single dose of ICS.

Effect of combination treatment (ICS and LABA)
Combination inhalers are going to be used as the first line treatment in asthma. Recently, it has been shown that combination treatment produced a clinically significant improvement in health status and the greatest reduction in daily symptoms in COPD. It is important, however, to monitor the underlying airway and alveolar inflammation in both diseases independently of patient's lung function and symptoms, which are affected by LABA. Surrogate markers may help us to see whether there is an additional anti-inflammatory effect of combination in these patients treatment in clinic.

There is evidence that symptom driven dosing with combination inhalers (inhaled steroid and long acting β_2 agonists) may be used in the future when the dose of the steroid could be determined by the degree of symptoms at a particular time. We suggest that the high sensitivity of exhaled NO may be used to adjust doses based on control of inflammation in asthma treatment. This is important as the long acting β_2 agonist may control symptoms and therefore mask the underlying inflammation which may not be adequately suppressed by corticosteroids. Portable, simple, and inexpensive exhaled NO analysers (based on measurements other than the chemiluminescence principle of NO detection) could be available in the next few years, making this approach feasible in the future.

Clinical research in biomarkers of asthma and COPD: needs and opportunities

Proteins, cytokines and metabolic markers. Novel technologies

Measurement and identification of proteins in exhaled condensate is controversial [4], although higher concentrations of total protein in exhaled condensate have been found in young smokers compared to non-smokers. Various proteins derived from airways and unlikely to be contaminated with saliva have been detected in EBC by two-dimensional electrophoresis. Although their range and source are still unclear, the proteins recovered in EBC might be used to non-invasively monitor respiratory diseases in the future.

Novel technologies, such as genomics, proteomics and metabonomics, which respectively can characterize the response of living systems to chemical exposure in terms of gene expression, protein expression, or metabolic regulation may help to speed up drug development by the pharmaceutical industry. These non-invasive technologies offer rapid, mechanistic information and are to some degree quantitative. They may facilitate incorporation of toxicological and clinical data at earlier stages of drug development with particular emphasis on biomarker discovery and characterization.

References

1 Kharitonov SA, Barnes PJ (2001) Exhaled markers of pulmonary disease. *Am J Respir Crit Care Med* 163: 1693–1722
2 Kharitonov SA, Barnes PJ (2002) Biomarkers of some pulmonary diseases in exhaled breath. *Biomarkers* 7: 1–32
3 Kharitonov SA, Barnes PJ (2001) Exhaled markers of inflammation. *Curr Opin Allergy Clin Immunol* 1: 217–224
4 Kharitonov SA, Barnes PJ (2000) Clinical aspects of exhaled nitric oxide. *Eur Respir J* 16: 781–792
5 Kharitonov SA, Donnelly LE, Montuschi P, Corradi M, Collins JV, Barnes PJ (2002) Dose-dependent onset and cessation of action of inhaled budesonide on exhaled nitric oxide and symptoms in mild asthma. *Thorax* 57: 889–896
6 Tsoukias NM, George SC (1998) A two-compartment model of pulmonary nitric oxide exchange dynamics. *J Appl Physiol* 85: 653–666
7 Sugiura H, Ichinose M, Yamagata S, Koarai A, Shirato K, Hattori T (2003) Correlation between change in pulmonary function and suppression of reactive nitrogen species production following steroid treatment in COPD. *Thorax* 58: 299–305
8 Fabbri LM, Romagnoli M, Corbetta L, Casoni G, Busljetic K, Turato G, Ligabue G, Ciaccia A, Saetta M, Papi A (2003) Differences in airway inflammation in patients with fixed airflow obstruction due to asthma or chronic obstructive pulmonary disease. *Am J Respir Crit Care Med* 167: 418–424
9 Kharitonov SA, Chung FK, Evans DJ, O'Connor BJ, Barnes PJ (1996) The elevated level of exhaled nitric oxide in asthmatic patients is mainly derived from the lower respiratory tract. *Am J Respir Crit Care Med* 153: 1773–1780
10 Brindicci C, Cosio B, Gajdocsi R, Collins JV, Bush A, Abdallah S, Barnes PJ (2002) Extended exhaled NO measurements at different exhalation flows may differentiate between bronchial and alveolar inflammation in patients with asthma and COPD. *Eur Respir J* 20: 174s
11 Paredi P, Kharitonov SA, Barnes PJ (2002) Analysis of expired air for oxidation products. *Am J Respir Crit Care Med* 166: S31–S37
12 Paredi P, Kharitonov SA, Leak D, Ward S, Cramer D, Barnes PJ (2000) Exhaled ethane, a marker of lipid peroxidation, is elevated in chronic obstructive pulmonary disease. *Am J Respir Crit Care Med* 162: 369–373
13 Corradi M, Montuschi P, Donnelly LE, Pesci A, Kharitonov SA, Barnes PJ (2001) Increased nitrosothiols in exhaled breath condensate in inflammatory airway diseases. *Am J Respir Crit Care Med* 163: 854–858
14 Montuschi P, Kharitonov SA, Carpagnano E, Culpitt SV, Russell R, Collins JV, Barnes PJ (2000) Exhaled prostaglandin E2: a new biomarker of airway inflammation in COPD. *Am J Respir Crit Care Med* 161: A821
15 Montuschi P, Barnes PJ (2002) Exhaled leukotrienes and prostaglandins in asthma. *J Allergy Clin Immunol* 109: 615–620

16   Montuschi P, Kharitonov SA, Ciabattoni G, Barnes PJ (2003) Exhaled leukotrienes and prostaglandins in COPD. *Thorax* 58: 585–588
17   Hanazawa T, Kharitonov SA, Barnes PJ (2000) Increased nitrotyrosine in exhaled breath condensate of patients with asthma. *Am J Respir Crit Care Med* 162: 1273–1276
18   Csoma Z, Kharitonov SA, Balint B, Bush A, Wilson NM, Barnes PJ (2002) Increased leukotrienes in exhaled breath condensate in childhood asthma. *Am J Respir Crit Care Med* 166: 1345–1349
19   Hanazawa T, Kharitonov SA, Oldfield W, Kay AB, Barnes PJ (2000) Nitrotyrosine and cystenyl leukotrienes in breath condensates are increased after withdrawal of steroid treatment in patients with asthma. *Am J Respir Crit Care Med* 161: A919
20   Biernacki WA, Kharitonov SA, Barnes PJ (2003) Increased leukotriene B4 and 8-isoprostane in exhaled breath condensate of patients with exacerbations of COPD. *Thorax* 58: 294–298
21   Kharitonov SA, Donnelly LE, Corradi M, Montuschi P, Barnes PJ (2000) Dose-dependent onset and duration of action of 100/400 mcg budesonide on exhaled nitric oxide and related changes in other potential markers of airway inflammation in mild asthma. *Am J Respir Crit Care Med* 161: A186
22   Montuschi P, Collins JV, Ciabattoni G, Lazzeri N, Corradi M, Kharitonov SA, Barnes PJ (2000) Exhaled 8-isoprostane as an *in vivo* biomarker of lung oxidative stress in patients with COPD and healthy smokers. *Am J Respir Crit Care Med* 162: 1175–1177
23   Paredi P, Ward S, Cramer D, Barnes PJ, Kharitonov SA (2003) A new method for the non-invasive measurement of bronchial blood flow. *Am J Respir Crit Care Med* 167: A448
24   Paredi P, Kharitonov SA, Barnes PJ (2002) Faster rise of exhaled breath temperature in asthma. A novel marker of airway inflammation? *Am J Respir Crit Care Med* 165: 181–184
25   Paredi P, Caramori G, Cramer D, Ward S, Ciaccia A, Papi A, Kharitonov SA, Barnes PJ (2003) Slower rise of exhaled breath temperature in chronic obstructive pulmonary disease. *Eur Respir J* 21: 439–443
26   Kharitonov SA (1999) Exhaled nitric oxide and carbon monoxide in asthma. *Eur Respir J* 9: 212–218
27   van Den Toorn LM, Prins JB, Overbeek SE, Hoogsteden HC, de Jongste JC (2000) Adolescents in clinical remission of atopic asthma have elevated exhaled nitric oxide levels and bronchial hyperresponsiveness. *Am J Respir Crit Care Med* 162: 953–957
28   Kharitonov SA, Gonio F, Kelly C, Meah S, Barnes PJ (2003) Reproducibility of exhaled nitric oxide measurements in healthy and asthmatic adults and children. *Eur Respir J* 21: 433–438
29   Yates DH, Kharitonov SA, Thomas PS, Barnes PJ (1996) Endogenous nitric oxide is decreased in asthmatic patients by an inhibitor of inducible nitric oxide synthase. *Am J Respir Crit Care Med* 154: 247–250
30   Hansel TT, Kharitonov SA, Donnelly LE, Erin EM, Currie MG, Moore WM, Manning PT, Recker DP, Barnes PJ (2003) A selective inhibitor of inducible nitric oxide synthase

inhibits exhaled breath nitric oxide in healthy volunteers and asthmatics. *FASEB J* 17: 1298–1300

31 Horvath I, Donnelly LE, Kiss A, Kharitonov SA, Lim S, Chung FK, Barnes PJ (1998) Combined use of exhaled hydrogen peroxide and nitric oxide in monitoring asthma. *Am J Respir Crit Care Med* 158: 1042–1046

32 Hanazawa T, Kharitonov SA, Barnes PJ (2000) Increased nitrotyrosine in exhaled breath condensate of patients with asthma. *Am J Respir Crit Care Med* 162: 1273–1276

33 Baraldi E, Ghiro L, Piovan V, Carraro S, Zacchello F, Zanconato S (2003) Safety and success of exhaled breath condensate collection in asthma. *Arch Dis Child* 88: 358–360

34 Kharitonov SA, Barnes PJ (2001) Does exhaled nitric oxide reflect asthma control? Yes, it does! *Am J Respir Crit Care Med* 164: 727–728

35 Kharitonov SA, Barnes PJ, O'Connor BJ (1996) Reduction in exhaled nitric oxide after a single dose of nebulised budesonide in patients with asthma. *Am J Respir Crit Care Med* 153: A799

36 Kharitonov SA, Yates DH, Barnes PJ (1996) Inhaled glucocorticoids decrease nitric oxide in exhaled air of asthmatic patients. *Am J Respir Crit Care Med* 153: 454–457

37 Horvath I, Donnelly LE, Kiss A, Paredi P, Kharitonov SA, Barnes PJ (1998) Elevated levels of exhaled carbon monoxide are associated with an increased expression of heme oxygenase-1 in airway macrophages in asthma: a new marker of oxidative stress. *Thorax* 53: 668–672

38 Brieva JL, Danta I, Wanner A (2000) Effect of an inhaled glucocorticosteroid on airway mucosal blood flow in mild asthma. *Am J Respir Crit Care Med* 161: 293–296

39 Kumar SD, Brieva JL, Danta I, Wanner A (2000) Transient effect of inhaled fluticasone on airway mucosal blood flow in subjects with and without asthma. *Am J Respir Crit Care Med* 161: 918–921

40 Kharitonov SA, Barnes PJ (2003) Nitric oxide, nitrotyrosine, and nitric oxide modulators in asthma and chronic obstructive pulmonary disease. *Curr Allergy Asthma Rep* 3: 121–129

41 Papi A, Romagnoli M, Baraldo S, Braccioni F, Guzzinati I, Saetta M, Ciaccia A, Fabbri LM (2000) Partial reversibility of airflow limitation and increased exhaled NO and sputum eosinophilia in chronic obstructive pulmonary disease. *Am J Respir Crit Care Med* 162: 1773–1777

42 Montuschi P, Corradi M, Ciabattoni G, Nightingale JA, Kharitonov SA, Barnes PJ (1999) Increased 8-isoprostane, a marker of oxidative stress, in exhaled condensate of asthma patients. *Am J Respir Crit Care Med* 160: 216–220

43 Maziak W, Loukides S, Culpitt SV, Sullivan P, Kharitonov SA, Barnes PJ (1998) Exhaled nitric oxide in chronic obstructive pulmonary disease. *Am J Respir Crit Care Med* 157: 998–1002

44 Montuschi P, Kharitonov SA, Barnes PJ (2001) Exhaled carbon monoxide and nitric oxide in COPD. *Chest* 120: 496–501

45 Corradi M, Majori M, Cacciani GC, Consigli GF, De'Munari E, Pesci A (1999)

Increased exhaled nitric oxide in patients with stable chronic obstructive pulmonary disease. *Thorax* 54: 572–575

46 Ferreira IM, Sandrini A, Zamel N, Balter M, Chapman KR (2003) Effects of inhaled fluticasone propionate on exhaled nitric oxide, functional exercise capacity and quality of life in stable patients with COPD. *Am J Respir Crit Care Med* 167: A317

47 Brindicci C, Gajdosci R, Barnes PJ, Kharitonov SA (2003) Flow-independent exhaled nitric oxide parameters in asthma and COPD. *Am J Respir Crit Care Med* 167: A221

48 Hogman M, Holmkvist T, Wegener T, Emtner M, Andersson M, Hedenstrom H, Merilainen P (2002) Extended NO analysis applied to patients with COPD, allergic asthma and allergic rhinitis. *Respir Med* 96: 24–30

49 Ferreira IM, Hazari MS, Gutierrez C, Zamel N, Chapman KR (2001) Exhaled nitric oxide and hydrogen peroxide in patients with chronic obstructive pulmonary disease: effects of inhaled beclomethasone. *Am J Respir Crit Care Med* 164: 1012–1015

50 Dekhuijzen PN, Aben KK, Dekker I, Aarts LP, Wielders PL, van Herwaarden CL, Bast A (1996) Increased exhalation of hydrogen peroxide in patients with stable and unstable chronic obstructive pulmonary disease. *Am J Respir Crit Care Med* 154: 813–816

51 Carpagnano GE, Kharitonov SA, Foschino-Barbaro MP, Resta O, Gramiccioni E, Barnes PJ (2003) Increased inflammatory markers in the exhaled breath condensate of cigarette smokers. *Eur Respir J* 21: 589–593

52 Corradi M, Rubinstein I, Andreoli R, Manini P, Caglieri A, Poli D, Alinovi R, Mutti A (2003) Aldehydes in exhaled breath condensate of patients with chronic obstructive pulmonary disease. *Am J Respir Crit Care Med* 167: 1380–1386

Systemic features of COPD

Alvar G.N. Agustí

Servei de Pneumología, Hospital Universitari Son Dureta, IUNICS, Andrea Doria 55, 07014 Palma de Mallorca, Spain

Introduction

Chronic obstructive pulmonary disease (COPD) is a disorder of the lung parenchyma characterised by varying combinations of chronic bronchitis and emphysema [1]. Only recently we have realized that COPD is also associated with significant abnormalities outside of the lungs, the so-called systemic features of COPD. These systemic features are clinically relevant because they contribute to jeopardize significantly the quality of life and the prognosis of the patient. Thus, they constitute clear therapeutic targets [2]. This chapter reviews currently accepted systemic features of COPD and anticipates some others of potential future relevance.

Systemic inflammation

The key pathogenic mechanism in COPD is an excessive/inadequate inflammatory response of the lungs to a variety of noxious inhaled gases or particles (mostly cigarette smoking), characterised by increased concentration of pro-inflammatory cytokines, augmented number of inflammatory cells, and evidence of oxidative stress [1]. It is less often realised, however, that similar inflammatory changes can also be detected in the systemic circulation of these patients. Yet, numerous studies have now reported increased levels of circulating cytokines and acute phase reactants [3–7], abnormal inflammatory cells [8–13] and oxidative stress [14, 15] in the peripheral circulation of patients with COPD.

Increased plasma levels of pro-inflammatory cytokines

Patients with clinically-stable COPD show increased concentrations of TNF-α, its receptors (TNFR-55 and TNFR-75), IL-6, IL-8, C-reactive protein, LPS binding protein, Fas and Fas-L [3–7]. Generally, these abnormalities were more pronounced during exacerbations of the disease [6]. The origin of systemic inflammation in COPD is unclear but several, not mutually exclusive, possibilities exist. First, tobacco smoke can cause by itself (i.e., in the absence of COPD) significant extra-pulmonary diseases [16]. Second, the pulmonary inflammatory process that characterizes COPD can also be a potential source of systemic inflammation. Inflammatory lung cells release inflammatory cytokines like TNF-α, IL-6, IL-1β, macrophage inflammatory protein-1α and GM-CSF, and increase their oxidant production when interacting with atmospheric particles, necrotic cells and other inflammatory mediators [17]. These pro-inflammatory mediators may reach the systemic circulation and/or contribute to the activation of inflammatory cells during their transit through the pulmonary circulation. Finally, it is also possible that the skeletal muscles themselves can also be a source of pro-inflammatory cytokines in COPD [18].

Circulating inflammatory cells

Several circulating inflammatory cells appear abnormal in COPD. Burnett et al. demonstrated that neutrophils harvested from COPD patients showed enhanced chemotaxis and extracellular proteolysis [8]. Noguera et al. reported that circulating neutrophils from COPD produced more ROS (respiratory burst) than those from non-smokers or healthy smokers, both under basal conditions and after stimulation *in vitro* [9]. These same authors showed that the expression of several surface adhesion molecules, particularly Mac-1 (CD11b), in circulating neutrophils was higher in patients with stable COPD than in healthy controls [10]. Interestingly, this difference disappeared during exacerbations of the disease, suggesting neutrophil sequestration in the pulmonary circulation during exacerbations [10]. Other abnormalities described in circulating neutrophils in COPD include the down-regulation of one type of G-proteins (Gαs) [10]. Gαs is involved in the intra-cellular signal transduction pathway linked to CD11b expression [10] and, also, in the control of the intra-cellular vesicular trafficking [19], the latter being relevant for the correct activation of NADPH oxidase, the enzyme which is eventually responsible of the respiratory burst in neutrophils [9]. It is then likely that Gαs can play a role in the regulation of some of the abnormalities described in circulating neutrophils in COPD, namely the increased expression of surface adhesion molecules [10] and the augmented respiratory burst [9].

Circulating lymphocytes have not been studied as much as circulating neutrophils in COPD, yet there are also some indications of abnormal lymphocyte function in these patients. Sauleda et al. showed that the activity of cytochrome oxidase, the terminal enzyme in the mitochondrial electron transport chain, was increased in circulating lymphocytes harvested from patients with stable COPD, as compared to healthy non-smoker controls [11]. This could also be detected in circulating lymphocytes of patients with other chronic inflammatory diseases, such as bronchial asthma and chronic arthritis, suggesting that it may be a non-specific marker of lymphocyte activation in chronic inflammatory diseases [12].

Finally, other authors have shown that peripheral monocytes harvested from patients with COPD are capable of producing more TNF-α when stimulated *in vitro* than those obtained from healthy controls [13]. This was particularly evident in patients with COPD and low body weight, suggesting that an excessive production of TNF-α by peripheral monocytes may play a role in the pathogenesis of weight loss in COPD [13].

Systemic oxidative stress

To assess whether patients with COPD had systemic oxidative stress, Rahman and co-workers determined the plasma Trolox-equivalent antioxidant capacity (TEAC) and the levels of products of lipid peroxidation in non-smokers, healthy smokers and COPD patients [14]. They found that both indices were significantly increased by smoking and COPD, particularly during exacerbations of the disease [14]. Praticò et al. also found that the urinary levels of isoprostane $F_{2\alpha}$-III, a stable prostaglandin isomer formed by ROS dependent peroxidation of arachidonic acid, were higher in patients with COPD than in healthy controls matched for age, sex and smoking habit [15]. Again, differences were more pronounced during exacerbations of the disease [15].

Unexplained weight loss

Unexplained weight loss occurs in about 50% of patients with severe COPD and chronic respiratory failure, but can also be seen in patients with mild to moderate disease [20]. It is mostly due to loss of skeletal muscle mass, loss of fat mass contributing to a lesser extent [20]. The mechanisms underlying this phenomenon are discussed below, (see "Skeletal muscle dysfunction"). Unexplained weight loss is a significant prognostic factor in COPD [21, 22]. Further, it is independent of the prognostic indicators that assess the degree of pulmonary dysfunction, such as FEV_1 or PaO_2 [21, 22]. Therefore, it identifies a new systemic domain of COPD not considered by the traditional measures of lung function. Because prognosis improves in those COPD patients who regain weight after appropriate therapy, in the absence of changes in lung function [21], unexplained weight loss has to be taken into account in the clinical assessment of patients with COPD [23].

Skeletal muscle dysfunction

Skeletal muscle dysfunction (SMD) is common in patients with COPD and contributes significantly to limit their exercise capacity and quality of life [2]. It is characterised by two different, but possibly related, phenomena: (1) net loss of muscle mass, an intrinsic muscular phenomenon; and (2) dysfunction of the remaining muscle. The latter may be secondary to either intrinsic muscle alterations (mitochondrial abnormalities, loss of contractile proteins) or alterations in the external milieu under which the muscle works (hypoxia, hypercapnia, acidosis), resulting from the abnormalities of pulmonary gas exchange that characterise COPD. Although conceptually important, the separation of these two aspects of SMD is extremely difficult *in vivo* and, probably, both play some role in any given patient. With this caveat in mind, the text that follows discusses potential mechanisms of SMD in COPD.

Sedentarism

Patients with COPD often adopt a sedentary lifestyle. Physical inactivity causes net loss of muscle mass, reduces the force generation capacity of the muscle and decreases its resistance to fatigue. Exercise training improves muscle function in COPD patients [24, 25], indicating that sedentarism is likely to be an important contributor to SMD. However, complete normalisation of muscle physiology is often not fully achieved after rehabilitation and, more importantly, some biochemical abnormalities found in the muscles are unlikely to be explained by physical inactivity [12]. For instance, at variance with the normal training response, exercise in patients with COPD enhances the release of amino-acids from skeletal muscle, particularly alanine and glutamine [26], suggesting the presence of intrinsic muscle abnormalities of the intermediate amino-acid metabolism [27].

Tissue hypoxia

Several observations support a potential pathogenic role for tissue hypoxia in the development of SMD in COPD. First, chronic hypoxia suppresses protein synthesis in muscle cells, causes net loss of amino acids and reduces the expression of myosin heavy chain isoforms [28]. Second, healthy subjects at high altitude (hypobaric hypoxia) lose muscle mass [29]. Third, skeletal muscle from patients with COPD and chronic respiratory failure present structural (decrease of type I fibers [30]) and functional alterations (up-regulation of mitochondrial cytochrome oxidase [12]) proportional to the severity of arterial hypoxemia.

Systemic inflammation

As discussed above, these patients show increased plasma levels of a variety of pro-inflammatory cytokines, particularly TNF-α [3–5, 7, 31]. TNF-α can affect muscle cells in several ways [32]. In differentiated myocites studied *in vitro*, TNF-α activates the transcription factor NF-κB and degrades myosin heavy chains through the ubiquitin-proteasome complex (U/P) [32]. Several studies have now shown that a disregulation of the U/P system contributes to the loss of muscular mass caused by sepsis or tumours in rats [33]. Whether this also occurs in COPD patients has not yet been investigated. Alternatively, TNF-α can induce the expression of a variety of genes, such as the inducible form of the nitric oxide synthase (iNOS) (see below), the TNF-α gene itself or those of many other pro-inflammatory cytokines, that would create a closed loop and contribute to the persistence and amplification of the inflammatory cascade [32]. Finally, TNF-α can induce apoptosis in several cell systems and we have shown recently that excessive apoptosis of skeletal muscle do indeed occur in patients with COPD and weight loss [34].

Oxidative stress

Systemic oxidative stress occurs in COPD [14, 15]. It can cause muscle fatigue [35] and facilitates proteolysis [33]. Further, the regulation of glutathione (GSH), the most important intra-cellular anti-oxidant [18], seems abnormal in skeletal muscle of patients with COPD [36].

Nitric oxide

The role of nitric oxide (NO) in the pathogenesis of SMD in COPD is unclear, but it could play a mechanistic role through several, not mutually exclusive, pathways. First, given that the number of capillaries in skeletal muscle of patients with COPD is lower than normal [37], it is conceivable that the expression of the endothelial isoform of the NO synthase (eNOS) will also be reduced. This can alter the control of the microcirculation and the supply of oxygen to the working muscle, resulting eventually in tissue hypoxia. And, second, systemic inflammation can up-regulate the expression of the inducible form of NOS (iNOS) in skeletal muscle, causing protein nitrotyrosination and facilitating protein degradation through the U/P system [33], enhancing skeletal muscle apoptosis [38] and/or causing contractile failure [39]. Results from our laboratory indicate that excessive protein nitrotyrosination and skeletal muscle apoptosis do both indeed occur in patients with COPD and low body weight [34, 40].

Tobacco smoke

Although it is accepted that tobacco smoke is the main risk factor for COPD [1], much less attention has been paid to the potential effects of tobacco smoke on SMD in COPD. Yet, tobacco smoke contains many substances potentially harmful for the skeletal muscle. For instance, nicotine alters the expression of important growth factors involved in the maintenance of the muscular mass, such as TGF-β1 [41], and competes with acetylcholine for its receptor at the neuromuscular junction, thus having the potential to affect directly muscle contraction [42].

Genetic background

Not all patients with COPD loose muscle mass during the course of their disease [20], suggesting a potential role for the genetic background of the individual. The potential genes involved in this process are unknown but the angiotensin converting enzyme (ACE) gene is a potential candidate because it is known to influence the muscle response to training in athletes [43] and the development of right ventricular hypertrophy in patients with COPD [44]. Further, a very recent report has shown that the use of ACE inhibitors can reduce the normal decline of muscle mass that occurs in ageing and can improve exercise capacity [45]. Other genes potentially involved in SMD include the transcription factors MyoD and MEF-2, as well as genes related to the histone acetylation-deacetylacion process (CBP/p300; HDAC-5), which have very recently been shown to play a fundamental role in the failure of muscle cells to regenerate after injury in patients with cancer cachexia [46, 47]. Their role in COPD has not been explored yet.

Endocrine disruption

Low testosterone and growth hormone levels [48, 49] and reduced plasma leptin concentration [50–52] are other potential mechanisms that, alone or in combination, can contribute to SMD in patients with COPD.

Other (potential) systemic effects of COPD

The text that follows discusses the potential role of other, still poorly defined, systemic features of potential relevance in COPD.

Cardiovascular effects

Using echo-doppler technology it has been shown that the endothelial function of the renal circulation is abnormal in patients with COPD [53, 54]. Whether or not this abnormality may also occur in other systemic vascular territories is not known at present. Likewise, in the absence of coronary artery disease and overt cor pulmonale, it is presently unclear whether left ventricular function is normal in stable patients with COPD. Cardiac output increases normally during exercise, even in severe COPD [55]. However, at peak exercise, cardiac output is about 50% of what a normal subject of the same age could achieve (at a correspondingly higher level of oxygen uptake, of course) [55]. This observation admits two potential explanations. First, despite the capacity of the heart to generate a higher cardiac output, this does not increase more than needed for the level of exercise. Second, left ventricular function may be compromised in COPD, and a higher cardiac output could not be achieved. This will require further studies.

Nervous system effects

Different aspects of the nervous system may be abnormal in patients with COPD. Using nuclear magnetic resonance (spectroscopy). Mathur et al. have recently shown that the bio-energetic metabolism of the brain is altered in these patients [56]. Whether this represents a physiological adaptation to chronic hypoxia, as it occurs at altitude [57], or it may be considered another systemic effect of COPD mediated by other unknown mechanisms is unclear. Likewise, patients with COPD present a high prevalence of depression [58]. Although this may simply represent a physiological response to a chronic debilitating disease, it is equally plausible that it may bear some relationship to the systemic inflammation that occurs in COPD, because TNF-α and other cytokines and molecules such as NO have been implicated in the pathogenesis of depression in several experimental models [59]. Finally, some recent data suggest that the autonomic nervous system may also be altered in patients with COPD [60]. Takabatake et al. showed indirect evidence of abnormal autonomic control in patients with COPD, particularly those with low body weight [60].

Osteo-skeletal effects

The prevalence of osteoporosis is increased in patients with COPD [61]. This can have multiple causes, including malnutrition, sedentarism, smoking, steroid treatment and systemic inflammation [61]. Thus, excessive osteoporosis in relation to age could also be considered a systemic effect of COPD [62]. In fact, it is interesting to note that emphysema and osteoporosis are both characterised by net loss of lung or bone tissue mass and, pictorially, an osteoporotic bone looks quite similar to an emphysematous lung! It is therefore tempting to speculate that, perhaps, both conditions share common mechanisms explaining the accelerated loss of tissue mass or its defective repair. This intriguing possibility merits future studies because a better understanding of the causes of the "excessive" osteoporosis of COPD may allow the design of new therapeutic alternatives that, eventually, may contribute to palliate its symptoms and to reduce the associated health care costs.

Conclusions

COPD can no longer be considered a disease affecting only the lungs. Available evidence indicates that COPD is associated to important systemic features. Further it is now clear that the clinical assessment of COPD ought to take into consideration the systemic components of the disease and that the treatment of these extra-pulmonary effects appear to be important in the clinical management of the disease. Therefore, a better understanding of the systemic features of COPD may allow the development of new therapeutic strategies that might eventually result in a better health status and better prognosis for the patients suffering this devastating disease.

Acknowledgements
Supported, in part, by Fondo de Investigación Sanitaria (RTIC C03/11), SEPAR and ABEMAR.

References

1   Pauwels RA, Buist AS, Calverley PM, Jenkins CR, Hurd SS (2001) Global strategy for the diagnosis, management, and prevention of chronic obstructive pulmonary disease. NHLBI/WHO Global Initiative for Chronic Obstructive Lung Disease (GOLD) Workshop summary. *Am J Respir Crit Care Med* 163: 1256–1276
2   American Thoracic Society, European Respiratory Society (1999) Skeletal muscle dysfunction in chronic obstructive pulmonary disease. *Am J Respir Crit Care Med* 159: S1–S40
3   Di Francia M, Barbier D, Mege JL, Orehek J (1994) Tumor necrosis factor-alpha levels and weight loss in chronic obstructive pulmonary disease. *Am J Respir Crit Care Med* 150: 1453–1455
4   Schols AM, Buurman WA, Staal-van den Brekel AJ, Dentener MA, Wouters EFM (1996) Evidence for a relation between metabolic derangements and increased levels of inflammatory mediators in a subgroup of patients with chronic obstructive pulmonary disease. *Thorax* 51: 819–824
5   Yasuda N, Gotoh K, Minatoguchi S, Asano N, Nishigaki K, Nomura M, Ohno A, Watanabe M, Sano H, Kumada H et al (1998) An increase of soluble Fas, an inhibitor of apoptosis, associated with progression of COPD. *Respir Med* 92: 993–999
6   Agustí AGN, Noguera A, Sauleda J, Miralles C, Batle S, Busquets X (2003) Systemic inflammation in chronic respiratory diseases. *Eur Respir Mon* 24: 46–55

7   Eid AA, Ionescu AA, Nixon LS, Lewis-Jenkins V, Matthews SB, Griffiths TL, Shale DJ (2001) Inflammatory response and body composition in chronic obstructive pulmonary disease. *Am J Respir Crit Care Med* 164: 1414–1418

8   Burnett D, Hill SL, Chamba A, Stockley RA (1987) Neutrophils from subjects with chronic obstructive lung disease show enhanced chemotaxis and extracellular proteolysis. *Lancet* 2: 1043–1046

9   Noguera A, Batle S, Miralles C, Iglesias J, Busquets X, MacNee W, Agustí AGN (2001) Enhanced neutrophil response in chronic obstructive pulmonary disease. *Thorax* 56: 432–437

10 Noguera A, Busquets X, Sauleda J, Villaverde JM, MacNee W, Agustí AGN (1998) Expression of adhesion molecules and G proteins in circulating neutrophils in chronic obstructive pulmonary disease. *Am J Respir Crit Care Med* 158: 1664–1668

11 Sauleda J, Garcia-Palmer FJ, Gonzalez G, Palou A, Agustí AGN (2000) The activity of cytochrome oxidase is increased in circulating lymphocytes of patients with chronic obstructive pulmonary disease, asthma, and chronic arthritis. *Am J Respir Crit Care Med* 161: 32–35

12 Sauleda J, García-Palmer F, Wiesner RJ, Tarraga S, Hanting I, Thomas P, Gomez C, Saus C, Palou A, Agustí AGN (1998) Cytochrome oxidase activity and mitochondrial gene expression in skeletal muscle of patients with chronic obstructive pulmonary disease. *Am J Respir Crit Care Med* 157: 1413–1417

13 De Godoy I, Donahoe M, Calhoun WJ, Mancino J, Rogers RM (1996) Elevated TNF-α production by peripheral blood monocytes of weight-losing COPD Patients. *Am J Respir Crit Care Med* 153: 633–637

14 Rahman I, Morrison D, Donaldson K, MacNee W (1996) Systemic oxidative stress in asthma, COPD, and smokers. *Am J Respir Crit Care Med* 154: 1055–1060

15 Praticò D, Basili S, Vieri M, Cordova C, Violi F, Fitzgerald GA (1998) Chronic obstructive pulmonary disease is associated with an increase in urinary levels of isoprostane $F_{2\alpha}$-III, an index of oxidant stress. *Am J Respir Crit Care Med* 158: 1709–1714

16 Raitakari OT, Adams MR, McCredie RJ, Griffiths KA, Celermajer DS (1999) Arterial endothelial dysfunction related to passive smoking is potentially reversible in healthy young adults. *Ann Intern Med* 130: 578–581

17 van Eeden SF, Tan WC, Suwa T, Mukae H, Terashima T, Fujii T, Qui D, Vincent R, Hogg JC (2001) Cytokines involved in the systemic inflammatory response induced by exposure to particulate matter air pollutants (PM_{10}) *Am J Respir Crit Care Med* 164: 826–830

18 Reid MB (2001) COPD as a muscle disease. *Am J Respir Crit Care Med* 164: 1101–1102

19 Zheng B, Ma YC, Ostrom RS, Lavoie C, Gill GN, Insel PA, Huang XY, Farquhar MG (2001) RGS-PX1, a GAP for GalphaS and sorting nexin in vesicular trafficking. *Science* 294: 1939–1942

20 Schols AMWJ, Soeters PB, Dingemans AMC, Mostert R, Frantzen PJ, Wouters EFM

(1993) Prevalence and characteristics of nutritional depletion in patients with stable COPD eligible for pulmonary rehabilitation. *Am Rev Respir Dis* 147: 1151–1156

21 Schols AM, Slangen J, Volovics L, Wouters EF (1998) Weight loss is a reversible factor in the prognosis of chronic obstructive pulmonary disease. *Am J Respir Crit Care Med* 157: 1791–1797

22 Landbo C, Prescott E, Lange P, Vestbo J, Almdal TP (1999) Prognostic value of nutritional status in chronic obstructive pulmonary disease. *Am J Respir Crit Care Med* 160: 1856–1861

23 Celli B, Cote C, Marin J, Montes de Oca M, Casanova C, Mendez MR (2000) The SCORE: a new COPD staging system combining 6MWD, MRD dyspnea, FEV$_1$ and PaO$_2$ as predictors of health care resources utilization (HCRU). *Am J Respir Crit Care Med* 161: A749

24 Maltais F, Leblanc P, Jobin J, Berube C, Bruneau J, Carrier L, Breton MJ, Falardeau G, Belleau R (1997) Intensity of training and physiologic adaptation in patients with chronic obstructive pulmonary disease. *Am J Respir Crit Care Med* 155: 555–561

25 Sala E, Roca J, Marrades RM, Alonso J, Gonzalez de Suso JM, Moreno A, Barbera JA, Nadal J, de Jover L, Rodriguez-Roisin, Wagner PD (1999) Effects of endurance training on skeletal muscle bioenergetics in chronic obstructive pulmonary disease. *Am J Respir Crit Care Med* 159: 1726–1734

26 Engelen MP, Wouters EF, Deutz NE, Does JD, Schols AM (2001) Effects of exercise on amino acid metabolism in patients with chronic obstructive pulmonary disease. *Am J Respir Crit Care Med* 163: 859–864

27 Engelen MP, Schols AM, Does JD, Gosker HR, Deutz NE, Wouters EF (2000) Exercise-induced lactate increase in relation to muscle substrates in patients with chronic obstructive pulmonary disease. *Am J Respir Crit Care Med* 162: 1697–1704

28 Bigard AX, Sanchez H, Birot O, Serrurier B (2000) Myosin heavy chain composition of skeletal muscles in young rats growing under hypobaric hypoxia conditions. *J. Appl. Physiol* 88: 479–486

29 Bigard AX, Brunet A, Guezennec CY, Monod H (1991) Skeletal muscle changes after endurance training at high altitude. *J Appl Physiol* 71: 2114–2121

30 Jakobsson P, Jorfeldt L, Brundin A (1990) Skeletal muscle metabolites and fibre types in patients with advanced chronic obstructive pulmonary disease (COPD), with and without chronic respiratory failure. *Eur Respir J* 3: 192–196

31 Sauleda J, Noguera A, Busquets X, Miralles C, Villaverde JM, Agustí AGN (1999) Systemic inflammation during exacerbations of chronic obstructive pulmonary disease. Lack of effect of steroid treatment. *Eur Respir J* 14: 359s

32 Li YP, Schwartz RJ, Waddell ID, Holloway BR, Reid MB (1998) Skeletal muscle myocytes undergo protein loss and reactive oxygen-mediated NF-kappaB activation in response to tumor necrosis factor α. *FASEB J* 12: 871–880

33 Mitch WE, Goldberg AL (1996) Mechanisms of muscle wasting. The role of the Ubiquitin-Proteasome pathway. *N Engl J Med* 335: 1897–1905

34 Agustí AGN, Sauleda J, Miralles C, Gomez C, Togores B, Sala E, Batle S, Busquets X

(2002) Skeletal muscle apoptosis and weight loss in chronic obstructive pulmonary disease. *Am J Respir Crit Care Med* 166: 485–489

35   Reid MB, Shoji T, Moody MR, Entman ML (1992) Reactive oxygen in skeletal muscle. II. Extracellular release of free radicals. *J Appl Physiol* 73: 1805–1809

36   Rabinovich RA, Ardite E, Troosters T, Carbo N, Alonso J, Gonzalez de Suso JM, Vilaro J, Barbera JA, Polo MF, Argiles JM et al (2001) Reduced muscle redox capacity after endurance training in patients with chronic obstructive pulmonary disease. *Am J Respir Crit Care Med* 164: 1114–1118

37   Whittom F, Jobin J, Simard PM, Leblanc P, Simard C, Bernard S, Belleau R, Maltais F (1998) Histochemical and morphological characteristics of the vastus lateralis muscle in patients with chronic obstructive pulmonary disease. *Med Sci Sports Exerc* 30: 1467–1474

38   Brüne B, Von Knethen A, Sandau KB (1998) Nitric oxide and its role in apoptosis. *Eur J Pharmacol* 351: 261–272

39   Lanone S, Mebazaa A, Heymes C, Henin D, Poderoso JJ, Panis Y, Zedda C, Billiar T, Payen D, Aubier M, Boczkowski J (2000) Muscular contractile failure in septic patients: role of the inducible nitric oxide synthase pathway. *Am J Respir Crit Care Med* 162: 2308–2315

40   Gari PG, Busquets X, Morlá M, Sauleda J, Agusti AG (2002) Up-regulation of the inducible form of the nitric oxide synthase and nitrotyrosine formation in skeletal muscle of COPD patients with cachexia. *Eur Respir J* 20: 211s

41   Cucina A, Sapienza P, Corvino V, Borrelli V, Mariani V, Randone B, Santoro D'Angelo L, Cavallaro A (2000) Nicotine-induced smooth muscle cell proliferation is mediated through bFGF and TGF-beta 1. *Surgery* 127: 316–322

42   Broal P (1992) Main features of structure and function. In: P Broal (ed): *The central nervous system*. Oxford University Press, New York, 550

43   Williams AG, Rayson MP, Jubb M, World M, Woods DR, Hayward M, Martin J, Humphries SE, Montgomery HE (2000) The ACE gene and muscle performance. *Nature* 403: 614–

44   van Suylen RJ, Wouters EF, Pennings HJ, Cheriex EC, van Pol PE, Ambergen AW, Vermelis AM, Daemen MJ (1999) The DD genotype of the angiotensin converting enzyme gene is negatively associated with right ventricular hypertrophy in male patients with chronic obstructive pulmonary disease. *Am J Respir Crit Care Med* 159: 1791–1795

45   Onder G, Penninx BW, Balkrishnan R, Fried LP, Chaves PH, Williamson J, Carter C, DiBari M, Guralnik JM, Pahor M (2002) Relation between use of angiotensin-converting enzyme inhibitors and muscle strength and physical function in older women: an observational study. *Lancet* 359: 926–930

46   Guttridge DC, Mayo MW, Madrid LV, Wang CY, Baldwin AS (2000) NF-kappaB-induced loss of MyoD messenger RNA: possible role in muscle decay and cachexia. *Science* 289: 2363–2366

47   Muscat GE, Dressel U (2000) Not a minute to waste. *Nat Med* 6: 1216–1217

48   Kamischke A, Kemper DE, Castel MA, Luthke M, Rolf C, Behre HM, Magnussen H,

Nieschlag E (1998) Testosterone levels in men with chronic obstructive pulmonary disease with or without glucocorticoid therapy. *Eur Respir J* 11: 41–45
49 Casaburi R (1998) Rationale for anabolic therapy to facilitate rehabilitation in chronic obstructive pulmonary disease. *Baillieres Clin Endocrinol Metab* 12: 407–418
50 Creutzberg EC, Wouters EF, Vanderhoven-Augustin IM, Dentener MA, Schols AM (2000) Disturbances in leptin metabolism are related to energy imbalance during acute exacerbations of chronic obstructive pulmonary disease. *Am J Respir Crit Care Med* 162: 1239–1245
51 Schols AM, Creutzberg EC, Buurman WA, Campfield LA, Saris WH, Wouters EF (1999) Plasma leptin is related to proinflammatory status and dietary intake in patients with chronic obstructive pulmonary disease. *Am J Respir Crit Care Med* 160: 1220–1226
52 Takabatake N, Nadamura H, Abe S, Hino T, Saito H, Yuki H, Kato S, Tomoike H (1999) Circulating leptin in patients with chronic obstructive pulmonary disease. *Am J Respir Crit Care Med* 159: 1215–1219
53 Howes TQ, Deane CR, Levin GE, Baudouin SV, Moxham J (1995) The effects of oxygen and dopamine on renal and aortic blood flow in chronic obstructive pulmonary disease with hypoxemia and hypercapnia. *Am J Respir Crit Care Med* 151: 378–383
54 Baudouin SV, Bott J, Ward A, Deane C, Moxham J (1992) Short-term effect of oxygen on renal haemodynamics in patients with hypoxaemic chronic obstructive airways disease. *Thorax* 47: 550–554
55 Agustí AGN, Cotes J, Wagner PD (1997) Responses to exercise in lung diseases. *Eur Respir Mon* 6: 32–50
56 Mathur R, Cox IJ, Oatridge A, Shephard DT, Shaw RJ, Taylor-Robinson SD (1999) Cerebral bioenergetics in stable chronic obstructive pulmonary disease. Am. J. Respir. Crit Care Med. 160: 1994–1999
57 Hochachka PW, Clark CM, Brown WD, Stanley C, Stone CK, Nickles RJ, Zhu GG, Allen PS, Holden JE (1994) The brain at high altitude: Hypometabolism as a defense against chronic hypoxia. *J Cereb Blood Flow Metab* 14: 671–679
58 Wagena EJ, Huibers MJ, Van Schayck CP (2001) Antidepressants in the treatment of patients with COPD: possible associations between smoking cigarettes, COPD and depression. *Thorax* 56: 587–588
59 Holden RJ, Pakula IS, Mooney PA (1998) An immunological model connecting the pathogenesis of stress, depression and carcinoma. *Med Hypotheses* 51: 309–314
60 Takabatake N, Nakamura H, Minamihaba O, Inage M, Inoue S, Kagaya S, Yamaki M, Tomoike H (2001) A novel pathophysiologic phenomenon in cachexic patients with chronic obstructive pulmonary disease: the relationship between the circadian rhythm of circulating leptin and the very low-frequency component of heart rate variability. *Am J Respir Crit Care Med* 163: 1314–1319
61 Gross NJ (2001) Extrapulmonary effects of chronic obstructive pulmonary disease. *Curr Opin Pulm Med* 7: 84–92
62 Engelen MP, Schols AM, Lamers RJ, Wouters EF (1999) Different patterns of chronic tissue wasting among patients with chronic obstructive pulmonary disease. *Clin Nutr* 18: 275–280

Pulmonary rehabilitation

Annemie M.W.J. Schols and Emiel F.M. Wouters

Department of Respiratory Medicine, University Hospital Maastricht, P.O. Box 5800, 6202 AZ Maastricht, The Netherlands

Introduction

Rehabilitation has been practised for several decades, but its application in respiratory disease is relatively recent. Although by definition chronic obstructive pulmonary disease (COPD) is a disease state characterised by the presence of a progressive and irreversible airflow obstruction, the primary treatment traditionally consists of pharmacological modulation of the airflow limitation by bronchodilating and anti-inflammatory agents. Despite symptomatic relief after optimal pharmacological intervention, most COPD patients still suffer from functional deficit. This minimal therapeutic approach is therefore no longer considered as acceptable in a state of the art management of COPD aimed at reduction of symptoms, improvement in health status and prevention and treatment of secondary complications [1, 2].

In 1974 a committee of the American College of Chest Physicians proposed the following definition of pulmonary rehabilitation: "An art of medical practice wherein an individually tailored, multidisciplinary program is formulated which through accurate diagnosis, therapy, emotional support and education stabilises or reverses both physio-pathological and psycho-pathological manifestations of pulmonary diseases and attempts to return the patient to the highest possible functional capacity allowed by his handicap and overall life situation" [3].

More recent definitions were formulated by the National Institute of Health (NIH) and by a task force of the European Respiratory Society (ERS). The NIH defined pulmonary rehabilitation as "a multidimensional continuum of services directed to persons with pulmonary disease and their families, usually by an interdisciplinary team of specialists, with the goal of achieving and maintaining the individual's maximum level of independence and functioning in the community" [4]. According to the ERS task force, pulmonary rehabilitation must be considered as "a process which systematically uses scientifically based diagnostic management and treatment options in order to achieve the optimal daily functioning and health related quality of life of individual patients suffering from impairment and disability due to chronic respiratory diseases as measured by clinically and/or physiologically relevant outcome measures" [5]. The statement of the American Thoracic Society (ATS) on pulmonary rehabilitation supports previous definitions by defining pulmonary rehabilitation as "a multidisciplinary program of care for patients with chronic respiratory impairment which is individually tailored and designed to optimise physical and social performance and autonomy of the patient" [6].

These definitions refer to a philosophical concept of rehabilitation as a process to restore an individual to the fullest medical, mental, emotional, social and vocational potential of its capability. This holistic approach of rehabilitation is based on the definition of health by the WHO as a state of complete physical, mental and social wellbeing. Present definitions are also based on the widely applied international classification model for impairments, disablement and handicaps (ICIDH) [7]. Impairment in this context specifically refers to the affected organ system. Disablement describes the decreased exercise capacity while handicap refers to the overall influence on the ability of the person to function in society. Based on irreversibility of the primary organ impairment, rehabilitation initially included the whole continuum of multidisciplinary services directed to manage the consequences of impairment and disablement on a patients' life. Recent insight into the underlying pathophysiology of local and systemic impairment in COPD directly linked to the level of disablement, allows a more precise definition of rehabilitation in the management of COPD, i.e., a continuum of multidisciplinary services, directed to modulate the impact of the disease process on a patient's life. Scientifically-based intervention strategies directed to modulate local and systemic impairment and related disablement, should be considered as non-pharmacological treatment of the disease. Based on careful and systematic diagnostic work-up, non-pharmacological treatment modalities should be an integrated part in the COPD management process. This approach fits with the view of the revised ICIDH-2 classification stating that healthy functioning and disablement form a continuum as outcomes of an interaction between a person's physical or mental condition and the physical and social environment in which they live [8].

Selection of candidates for rehabilitation

Although pulmonary rehabilitation programs have been widely applied in COPD patients, it is unclear what patients are good candidates. The ATS statement considers pulmonary rehabilitation indicated for patients with chronic impairment who despite optimal medical management are dyspnoeic, have reduced exercise tolerance or experience a restriction in activities [6]. Several studies have clearly shown that disease severity in COPD should not only be based on FEV_1 but also on other factors of local impairment as well as on measures of disablement (i.e., maximal oxygen uptake, exercise capacity, pulmonary hyperinflation, maximal respiratory pressures, dyspnoea and quality of life ratings [9–11]).

Irrespective of the complexity of the underlying pathophysiology, improvements in dyspnoea and exercise tolerance are considered as attainable targets of intervention in pulmonary rehabilitation. Therefore, pulmonary rehabilitation refers to the whole spectrum of non-pharmacological treatment interventions, directed to attenuate the persistent impaired health status after present pharmacological treatment and should be considered in every COPD patient, independent of the degree of airflow limitation [12]. Indeed, similar gains in physical performance and health status were reported in COPD patients with severe, moderate and mild disease based on spirometric criteria (ATS) [12].

Dimensions of pulmonary rehabilitation programs

Based on the historically defined concept of pulmonary rehabilitation, each patient enrolled in a rehabilitation program has to be considered as a unique individual with specific physical and psychological impairment caused by the underlying disease. Therefore, pulmonary rehabilitation programs incorporate many different therapeutic modalities in a comprehensive, multidisciplinary care program. The efficacy and effectiveness of some components is scientifically proven. In order to improve quality of life, or to promote self-management behavior, it is also important to consider the different dimensions of pulmonary rehabilitation, the aim, the focus and the directness of the intervention [13]. The aim may range from reduction and control of respiratory symptoms, improvement in quality of life to reduction of psychological impact of physical impairment and disability. The focus can be on the individual, a group or the environment. The approach can be direct or indirect. Interventions that aim at improvement of, for example quality of life, will focus on improvement of general psychological, social, practical and physical well-being of the patient. Such interventions can involve physical exercise programs as well as stress-management programs, social skills training or different kinds of counseling and support. The level of focussing of the intervention depends on the aim of the intervention and the expected efficiency. Group training is highly appreciated by patients; psychological group interventions directed at both patients and partners can increase efficiency in order to achieve management goals. Furthermore, interventions can be directed at change or adaptation of the environment of the COPD patient. These interventions are often specified by the term "social engineering", because they try to modify living, work, or leisure-time situations and healthy lifestyles of the patient from a social or patient perspective [14]. Finally, the directness of an intervention has to be considered. Indirect interventions can be considered to improve social support for the patient or to train other professionals in intervention skills. Components of a rehabilitation program are individualized based on a careful assessment of the patient, not limited to lung function testing, but addressing physical and emotional deficits, knowledge of the disease, cognitive and psychosocial functioning as well as nutritional assessment. This assessment should not be performed once, but be an on-going process during rehabilitation. This theoretical approach however is still largely unattainable in most rehabilitation programs, due to limited resources for non-pharmacological intervention strategies in COPD.

Outcome of rehabilitation in COPD

Several studies documented the outcome of comprehensive pulmonary rehabilitation programmes. Ries et al. [15] compared the effects of a comprehensive pulmonary rehabilitation program including exercise reconditioning with education alone on physiological and psychosocial outcomes in COPD. Pulmonary rehabilitation consisted of twelve four-hour sessions of education, physical and respiratory care instruction, psychosocial support and supervised exercise training, followed by monthly reinforcement sessions for one year. The education group received two-hours sessions that included videotapes, lectures and discussions. The comprehensive rehabilitation program produced a significant increase in maximal exercise tolerance, maximal oxygen uptake, exercise endurance, self-efficacy for walking associated with a marked reduction of perceived breathlessness, muscle fatigue and shortness of breath. Most of these effects persisted for 18–24 months although benefits tended to diminish after one year. Positive effects of rehabilitation on dyspnoea were confirmed by the results of O'Donnell et al. [16], who demonstrated that supervised multi-modality endurance exercise training relieved both chronic and acute activity-related breathlessness and that this relief of breathlessness was related to a fall in ventilatory demand during exercise as a result of enhanced mechanical efficiency. This improvement in breathlessness was translated into significant improvement in exercise capacity and into greater ability to participate in activities of daily living. Goldstein et al. [17] reported similar results in a prospective randomized controlled trial of respiratory rehabilitation including 89 subjects. Exercise activities consisted of interval training, treadmill, upper-extremity training and leisure walking as part of an eight-weeks in-patient rehabilitation program. Significant improvements in exercise tolerance, measured by submaximal cycle time and walking distance were demonstrated that sustained for at least six months. Significant improvements were also noted in the scores on a dyspnoea questionnaire. Other studies demonstrated that beneficial effects are even achieved after home-based pulmonary rehabilitation programmes: improvements in maximal workload, symptom-limited oxygen uptake and maximal inspiratory pressure together with a decrease of lactate, inspiratory muscle load and dyspnoea during maximal exercise were reported. These effects were maintained over 18 months [18, 19]. Wedzicha et al. [20] tested the hypothesis that severity of respiratory disease affects the outcome of pulmonary rehabilitation. In a randomized, controlled study patients with

COPD were stratified for dyspnoea using the Medical Research Council (MRC) dyspnoea score and the patients were randomly assigned to an eight-week program of either exercise plus education or education alone. Improvements in exercise performance and health status were higher in patients with moderate levels of dyspnoea.

Others have confirmed positive effects of pulmonary rehabilitation on the short-term and long-term even by inexpensive, comprehensive out-patient programs [21–23], although continuation of supervised training is generally recommended in order to maintain training effects. A meta-analysis of respiratory rehabilitation demonstrated that pulmonary rehabilitation relieves dyspnoea and improves the control over COPD; these improvements were considered clinically relevant while the value of improvement in exercise capacity was less convincing [24].

More recent studies evaluated the outcome of pulmonary rehabilitation, in terms of cost-effectiveness. Goldstein et al. [25] reported an economic analysis of two months of in-patient rehabilitation followed by four months of out-patient supervision. The incremental cost of achieving improvements beyond the minimal clinically important difference in dyspnoea, emotional function and mastery was $ 11.597 (Canadian). More than 90% of this cost was attributable to the in-patient part of the program. Of the non-physician health care professionals, nursing was identified as the largest cost center, followed by physical therapy and occupational therapy.

Troosters et al. [26] reported that a six-month out-patient rehabilitation program that involved moderate-to-high training intensity did not alter pulmonary function, but did improve functional and maximal exercise performance, peripheral and respiratory muscle strength, and quality of life when compared to usual care in patients with severe COPD. These improvements in functional and maximal exercise performance were clinically relevant and were maintained 18 months after the onset of training. This out-patient program had a mean cost per patient of approximately $ 2,600 to achieve a mean improvement of 52 m in six-minute walking distance at six months. Griffiths et al. [27] analysed the effects of out-patient pulmonary rehabilitation on use of health care and patients' wellbeing over one year. They reported no difference between the rehabilitation and control groups in the number of patients admitted to the hospital but showed a difference in the number of days spent in hospital.

Furthermore, they demonstrated that the rehabilitation group had more primary care consultations at the general practitioners' premises than did the control group but fewer primary care home visits. The rehabilitation group showed greater improvements in walking ability and in health status. Such benefits in health status as well as in hospitalisations persisting for a period of two years after out-patient rehabilitation, are confirmed in literature [28].

In summary, there is now convincing evidence in literature for the efficacy and effectiveness of comprehensive pulmonary rehabilitation programs.

Components of rehabilitation programs

Exercise training

Impaired exercise tolerance is a prominent feature in patients suffering from COPD. Exercise limitation may be the result of changes in a wide spectrum of local impairment: reduced expiratory airflow as a consequence of poor elastic recoil; increased airways resistance leading to increased work of breathing and increased ventilatory drive; reduced pulmonary vascular bed and increased pulmonary vascular resistance contributing to exercise induced hypoxaemia; impaired cardiac output by impediment of right heart filling and left ventricular systolic function. Besides, leg fatigue due to peripheral muscle weakness is a common limiting symptom during exercise in COPD [29]. Several factors may contribute to skeletal muscle dysfunction in COPD: chronic inactivity, systemic inflammation, systemic corticosteroid administration, hypoxaemia, electrolyte disturbances and nutritional depletion. In addition intrinsic abnormalities in skeletal muscle structure and metabolism have been reported: decreased oxidative capacity [30], a greater proportion of fatigue-susceptible fibers as a consequence of shifts from type 1 fibers to type 2 fibers [31] and depletion of energy rich substrates [32]. A decreased oxidative capacity is related to an increased lactic acidosis for a given exercise work rate and enhanced ventilatory needs by increasing non-aerobic carbon dioxide production. This requirement imposes an additional burden on the respiratory muscles that already face increased impedance to breathing. Exercise in COPD may also induce early-onset of muscle intracellular acidosis [33]. Intrinsic muscular abnormalities are not comparable for lower limb muscle and the diaphragm. In severe COPD patients, opposite changes in diaphragmatic fiber composition have been reported towards a higher proportion of fatigue resistant fibers [34]. Mechanical disadvantage and altered muscle fiber length as a consequence of static and dynamic hyperinflation therefore mainly contribute to respiratory muscle dysfunction. A possible imbalance between inspiratory muscle function and increased muscle demand related to the increased resistive and elastic load is an important determinant of dyspnoea, susceptibility to inspiratory muscle fatigue, drive on the respiratory muscles and hypercapnia [35].

COPD specific abnormalities in respiratory and peripheral skeletal muscle function are generally not considered in the prescription of exercise training for COPD. However in order to determine the nature of exercise limitation and to be able to prescribe a tailor made exercise program appropriate exercise testing is a prerequisite.

Physiological outcome of exercise training in COPD
Although exercise training is considered as the cornerstone of a rehabilitation program, the physiological benefits of exercise training remained unclear until the

1990s. Due to their ventilatory limitation it was generally thought that COPD patients are unable to achieve a training intensity sufficiently high to train exercising muscles. Casaburi et al. [36] however clearly showed that physiologic training responses could be observed in these patients. At a given level of exercise, significant reductions in blood lactate, CO_2 production, minute ventilation, O_2 consumption and heart rate were observed. The ventilatory requirement for exercise dropped after an effective training program in proportion to the drop in blood lactate at a given work load. Based on these data and the results of other studies [37] it can be concluded that physiologic adaptation to training may occur in COPD patients. A reduction in lactic acid production by the contracting muscles is probably the main mechanism underlying this adaptive process. Indeed early lactic acid production during exercise is reported in COPD patients probably related to a decreased oxidative capacity and altered muscle substrate metabolism. Maltais et al. showed a decreased capacity of the Krebs cycle enzyme citrate synthase in the m. vastus lateralis in COPD [30] while Engelen demonstrated a relationship between decreased muscular glutamate status and early lactic acid production [38]. Subsequently an improvement in citrate synthase was shown [37] after a three month endurance training program being related to the reduction in exercise-induced lactic acidosis in these patients.

Beneficial effects of training in patients with COPD are also reported on skeletal muscle bioenergetics assessed by NMR-spectroscopy. The half-time of phosphocreatine (PCr) recovery fell significantly after an eight-weeks endurance training program and at a given submaximal work rate, improved bioenergetics was reflected in a decreased inorganic phosphate to phosphocreatine ratio and an increased intracellular pH. In summary, these data indicate that physiological changes provoked by endurance training are observed at the level of skeletal muscle adaptations during submaximal exercise [39].

Intensity, specificity and duration of lower extremity training
The optimal mode of exercise training still remains a matter of debate. In general, exercise training can be divided into two types: aerobic or endurance training and strength training. The majority of the studies of exercise training in COPD have focused on endurance training. In healthy subjects recommendations are available about duration, intensity and frequency for aerobic training [40, 41]. According to these recommendations, aerobic training calls for rhythmical, dynamic activity of large muscles, performed 3–4 times a week for 20–30 min per session at an intensity of at least 50% of maximal oxygen consumption. Such a program of aerobic training is capable of inducing structural and physiological adaptations that provide the trained individual with improved endurance for high-intensity activity. Most of the rehabilitation programs include exercise sessions of at least 30 min, 3–5 times a week. Although no ideal duration has been established, duration in

many programs is between eight and 12 weeks. In order to assess the optimal duration of a pulmonary rehabilitation program, one randomized controlled trial investigated a seven week twice-weekly out-patient based program with a comparable but shortened four week program. The seven weeks course of pulmonary rehabilitation provided greater benefits in terms of improvement in health status [42].

Limited information is also available on physiological outcome of different types of exercise testing. Most studies have investigated the physiological response of continuous training at a given work load in order to stress oxidative metabolism. Otherwise, interval training has been evaluated since by alternating high and lower training load it resembles more closely daily activity pattern especially in severe COPD patients. Furthermore interval training may also stress glycolytic metabolism. One comparative study indeed showed that continuous training resulted in a significant increase in oxygen consumption, and a decrease in minute ventilation and ventilatory equivalent for carbon dioxide at peak exercise capacity, while no changes in these measures were observed after interval training [43]. A significant reduction in lactic acid production was observed after both training modalities but most pronounced in the continuous training group. Remarkably, in the interval training group a decrease in leg pain was reported as well as a significant increase in peak workload.

Limited data is available comparing concentric exercise (positive work) with eccentric exercise (negative work) as part of the rehabilitation program. In patients with COPD the ventilatory requirements of eccentric exercise are considerably lower than those of concentric exercise at similar work loads resulting in a greater ventilatory reserve and less disturbed gas exchange [44]. Therefore, eccentric work might be a suitable type of exercise and training in patients with limited ventilatory reserves. In one randomized trial, the effects of eccentric exercise training in addition to general exercise training on exercise performance and health status were compared. It was reported that pulmonary rehabilitation improved exercise performance and health status similarly in both training groups but physiological training effects were observed only in the eccentrically trained group [45].

The optimal training intensity for COPD patients is still a matter of debate. In healthy subjects training is normally targeted on base of maximal heart rate (60–90% of predicted) or the percentage of maximal oxygen uptake (50–80% predicted) [40]. However, principles of exercise intensity derived from health subjects may not be applicable for pulmonary patients who are limited by breathing capacity and dyspnoea. Some investigators have reported that patients with COPD can tolerate high-intensity training. These patients could even be trained at an intensity which represents a higher percentage of maximum exercise tolerance than recommended for healthy subjects, because they can sustain ventilation at a high percentage of their maximum breathing capacity [46–48]. In some studies, it was even concluded that high intensity training is superior to low intensity training [36]. Others in contrast concluded that most patients with COPD are unable to achieve high intensity training, defined as a training intensity of 80% of baseline maximal power output [49]. Furthermore, these authors demonstrated that the intensity of training achieved, is not influenced by the initial baseline maximal oxygen consumption, age or the degree of airflow limitation.

Clark et al. [50] investigated the physiological benefits of an exercise program that concentrated on isolated conditioning of peripheral skeletal muscles rather than whole-body aerobic training on the premise that a cumulative set of individual limb exercises would be better tolerated by the patients than whole body exercise. Indeed the training group showed significant improvement in a variety of measures of upper and lower limb muscle performance, with no additional breathlessness. Furthermore, the training group showed a reduction in ventilatory equivalents for oxygen and carbon dioxide, both at peak exercise and at equivalent work rate. An increase in efficiency of peripheral oxygen extraction by the skeletal muscle was hypothesized to explain these physiological responses.

Vallet et al. [51] tested the effect of two methods of training, one individualized at the heart rate corresponding to the gas exchange threshold and the other at 50% of maximal heart reserve. While a significant increase in symptom-limited oxygen uptake and maximal oxygen pulse was measured after the individualized protocol, no significant changes were observed after the standard training protocol. Individualized training also exhibited a concomitant and gradual decrease in minute ventilation, carbon dioxide production and venous lactate concentration. These results clearly indicate that despite an apparently similar overall target training level, individualized training clearly optimized the physiological training effects in patients with chronic airflow limitation and decreased their ventilatory requirement.

Only limited data are available on the effects of strength training in patients with pulmonary disease. Strength training involves the performance of explosive tasks such as weightlifting over a short period of time. Simpson et al. [52] reported a 73% increase in cycling endurance time at 80% of maximal power output following eight weeks of weightlifting training of the upper and lower extremity muscles. Otherwise, no significant changes in maximal cycling exercise capacity or walking distance were observed. Others confirmed that weight training can improve treadmill walking endurance of patients with mild COPD and that this improvement in treadmill endurance correlated with improvements in upper and lower limb isokinetic sustained muscle strength following training [53]. The outcome of a combination of strength training and endurance training also needs further evaluation. In one study a combination of aerobic endurance training and strength training resulted in a significant increase in quadriceps strength, thigh muscle cross-sectional area and pectoralis major muscle strength but no specific influence on peak work rate, walking distance or health status was seen [54].

Upper extremity training

Patients with COPD frequently report disabling dyspnoea for daily activities involving the upper extremities such as combing the hairs, brushing teeth or shaving. It is known that even in healthy persons arm exercise is relatively more demanding than leg exercise. Some studies have demonstrated that arm elevation is related to a disproportionate increase in the diaphragmatic contribution to the generation of ventilatory pressures [55] and that arm elevation is a fatiguing task for the muscles involved as assessed by electro-myographic data. In COPD patients, studies have reported that arm exercise has effects on breathing pattern, recruitment of expiratory muscles as well as on the metabolic and ventilatory response pattern [56–60]. However, relatively few data have evaluated the effects of upper extremity (UE) training relative to lower extremity training. Studies have demonstrated that UE training leads to improved arm muscle endurance during isotonic arm ergometry [61] and that arm training conducted during a pulmonary rehabilitation program led to a reduced metabolic demand for arm exercise [55]. Based on present findings, it can be concluded that strength and endurance training of the UE improves arm function and that these exercises can be safely incorporated in rehabilitation programs for patients with pulmonary diseases. Further studies are needed to explore the effects of arm training on functional outcomes, to evaluate different forms of arm exercise training programs and to determine the effect of arm exercise training on respiratory muscle function.

Ventilatory muscle training

There is accumulating evidence in literature for respiratory muscle dysfunction especially in COPD patients. Four main factors may explain inspiratory muscle dysfunction in COPD: 1) mechanical disadvantage associated with hyperinflation; 2) altered muscle fiber length as important determinant of the force-generating capacity; 3) alterations in the intrinsic muscle structure, manifested by changes in fiber type composition and muscle mass; and 4) electrolyte disturbances [62]. Interventions directed to improve respiratory muscle performance have to consider the force developed during contraction as a fraction of the maximal force (measured by the ratio P_{breath}/P_{max}) and the duty cycle for the inspiratory muscles. Changes in duty cycle are difficult to obtain. Therefore, interventions directed to improve respiratory muscle performance focus on lowering the ratio P_{breath}/P_{MAX} by reducing the load on the respiratory muscles or by improving their force-generating capacity. Ventilatory muscle training is generally practiced in order to increase respiratory muscle strength. Two types of training are commonly applied in ventilatory muscle training: normocapnic sustained hyperpnoea and inspiratory resistance breathing. During normocapnic hyperpnoea, supernormal target ventilation is required for 15 to 20 minutes, during which carbon dioxide tension is kept constant. This form of training therefore requires complicated equipment to monitor the patients and requires a medical facility in order to train the patients. Inspiratory resistance training uses small hand-held devices based on a resistance, flow-dependent system or by applying a threshold valve as a flow-independent device.

A recent meta-analysis reviewed the effects of inspiratory resistive training [63]. The effects of 17 reviewed studies were rather disappointing: non-significant changes in Pi-max (11 studies) and in respiratory muscle endurance (nine studies) were reported. These findings demonstrate that control of the training stimulus may be exceedingly important to induce the expected physiologic training response. Other studies have now demonstrated that respiratory muscle training, if properly applied, results in improved respiratory muscle strength or endurance [64]. This improvement in respiratory muscle function was associated with a decreased sensation of dyspnoea. In patients that are ventilatory limited during exercise, target-flow inspiratory muscle training combined with peripheral muscle exercise training allowed for an additional improvement of walking distance and maximal exercise capacity compared to exercise training alone [65, 66], but these results could not be reproduced in patients with preserved inspiratory muscle function [67, 68].

Future studies on skeletal muscle training of ventilatory muscles relative to upper and lower limb muscles should also consider striking differences in muscular alterations between lower limb and respiratory muscles in COPD patients, probably reflecting the degree of muscle activity and muscle load.

Education

Patient education is generally used as "umbrella" term for various forms of goal-directed and systematically applied communication processes, directed at improvement of cognition, understanding and motivation, as well as on improvement of action- and decision-making possibilities of a patient to improve the coping with and recovery of the disease [69]. Ideally, patient education is a "planned learning experience using a combination of methods such as teaching, counseling and behavior modification techniques in order to influence patient knowledge and health behavior" [70]. Promotion of self-management behavior in COPD can be directed to improve adherence to medical advice with respect to medication and healthy lifestyle, aim at stabilization or retardation of progression or at avoidance of undesirable consequences and complications. Medical advice to chronically ill patients can also be directed at various aspects of cognition and behavior [14].

An optimal education program for patients with COPD can be formulated as follows [14]:

1. The program should be conducted by experts specially trained in techniques to change behavior or irrational cognitions;
2. Information should be provided in a structured way;

3. Although a group-program is preferable from a health economical perspective, a combination of an individualized program and a group-program may be most effective;
4. Both participation of the social environment and attention for the problems of the partners should have a high priority to maintain newly-acquired skills and cognitions in the home situation;
5. Both medical and psychosocial parameters should be emphasized;
6. The responsibility of the patient for his own health must be emphasized;
7. In order to promote the patient's self-activity and to support the maintenance of behavioral changes in the home situation, additional materials should be made available to be used at home;
8. Follow-up sessions are necessary to support the patient and his or her partner in the home situation;
9. Specific patient education interventions should be implemented in a multi-disciplinary program to improve physical and psychological functioning;
10. Short- and long-term effects have to be evaluated by valid measurements.

Studies concerning patient education in COPD patients are limited [71–75]. Most of the reported studies tried to improve self-management, decrease medical consumption, decrease life-stress and increase social support quality of life. Various professionals conducted most educational programs, but the environment of the patient was generally not involved and most of the studies did not pay attention to the partners. The overall impression in most studies is that some aspects of the disease are positively affected like depression, anxiety, optimism, wellbeing, the number and length of hospital admissions and use of health services. Most of these studies however report only short-term effects. Van den Broek reported the effects of a patient education group intervention program as part of a pulmonary rehabilitation program [14]. Patients were randomly assigned to an experimental group and a control group. Partners also participated to the study. Patients in the control group received medical advice and standard clinical care. The experimental group followed a structured educational program consisting of two components: an informative part and an educational part. The total program was directed at teaching self-management skills. Patients were followed for 12 months after the end of the rehabilitation program. Limited or no effects could be demonstrated on variables for psychological functioning, physical functioning or for social and practical functioning.

Stabilization or reversal of disease-related psychopathology was one of the initially defined goals of pulmonary rehabilitation. Personality traits and intra-psychic conflicts as well as acute psychological states as panic, anxiety or depression are widely recognized problem categories in patients with COPD. Specific psychosocial intervention strategies are usually required in order to modify these problems. Kaptein and Dekker recently reviewed the nature of psychosocial support in different rehabilitation programs. They concluded that relaxation techniques as predominantly passive form of intervention aimed at more controlled and efficient breathing, was the most frequently applied type of psychosocial support [76]. The authors concluded that future research is needed to assess the outcome of more specific psychosocial intervention strategies as well as to delineate the contribution of psychosocial intervention itself over and above pulmonary rehabilitation programs.

Nutritional support

The association between underweight and increased mortality risk has been well established in numerous retrospective studies ranging from selected COPD patients to population-based samples [1–3]. Two prospective studies even showed in COPD patients with a body mass index below 25 kg/m^2 that weight gain was associated with decreased mortality risk [2, 4]. It is not fully understood why COPD patients become underweight but weight loss and specifically loss of fat mass, is generally the result of a negative energy balance and appears to be more prevalent in patients with emphysema [5]. In contrast to an adaptive decreased energy metabolism during (semi) starvation, increased resting energy requirements have been observed in part of the COPD patients, linked to low-grade systemic inflammation [6, 7]. Studies in other chronic wasting diseases characterized by hypermetabolism and systemic inflammation (e.g., cancer, chronic heart failure, AIDS) have shown an adaptive decrease in activity induced energy expenditure so that daily energy expenditure is normal. In contrast, elevated activity induced and daily energy expenditure has been measured in free-living ambulatory COPD patients [8]. The cause of this disease-specific increase is not yet clear, but could be related to decreased mechanical efficiency of leg exercise through less efficient muscle energy metabolism, and increased oxygen cost of respiratory muscle activity due to lung hyperinflation. An obvious choice to improve energy balance might thus be to decrease energy expenditure. However, according to the recent GOLD guidelines, pulmonary rehabilitation is evidence based and exercise training a key intervention to improve limited functional abilities and maintain an active lifestyle [9]. Since COPD patients may have an elevated energy metabolism and should at the same time be advised to increase exercise, restricting energy output will be hard to realise and may not be desirable. This implies that COPD patients who suffer from weight loss, and even some weight stable patients, should be encouraged to increase their apparently normal energy intake. This could avoid weight loss, specific loss of muscle mass, and a related decrease in functional ability, or could help them regain weight. Besides optimising the treatment of patients who are already underweight, it is therefore important to detect and reverse involuntary weight loss in order to avoid functional decline. This may be achieved by increasing dietary intake *per se* or by altering dietary habits to include different (energy-dense) foods and optimum timing of meals/snacks in relation to symptoms and activity patterns.

Substrate oxidation and ventilation are intrinsically related and theoretically meal related dyspnoea and impaired ventilatory reserves might restrict the caloric amount and specifically the carbohydrate content of nutritional support in respiratory disease. Earlier studies indeed showed adverse effects of a carbohydrate rich energy overload on CO_2 production and exercise capacity [10], but these results were not confirmed when using a normal energy load [11]. In fact even positive effects of a carbohydrate rich supplement relative to a fat-rich supplement on lung function and dyspnoea sensation were reported [12].

Caloric supplementation versus dietary change

Nutritional interventions for COPD patients have focussed mainly on therapeutic caloric support, but few well-controlled studies (i.e., randomised controlled trials, RCTs) have so far been conducted. Ferreira and colleagues have recently reviewed the available studies on therapeutic dietary supplementation in a meta-analysis [13]. They managed to select only six RCTs that were considered to be of sufficient quality, of which two were double-blinded. The pooled effects, based on the analysis of a total of 277 subjects, as well as the results of the individual studies, showed that the effect of nutritional support on anthropometrics was minor at best and generally did not achieve clinical importance or statistical significance. Five of these studies used oral supplementation and four were conducted among out-patients. These results contradict to some extent the results reported by Baldwin and colleagues [14], who reviewed the literature on dietary advice and supplementation interventions for patients with disease-related malnutrition in general (including COPD). Their conclusion was that dietary supplementation resulted in better effects on body weight than dietary advice. The review by Ferreira and colleagues' made no distinction between what may be called "failure to intervene" on the one hand and "failure of the intervention" on the other. In some of the papers on which the meta-analysis was based, patients took the prescribed dietary supplements to replace regular meals instead of as additional calorie input. In such cases the intervention did not succeed in a relevant increase in energy intake and therefore no weight gain could be expected. In the studies that did accomplish to increase energy intake, functional improvements were also observed. Further, studies investigating the effect of dietary supplementation were often conducted among severe COPD cases, in whom besides a negative energy balance, also a specific negative protein balance is often observed (as discussed below). Nevertheless, the meta-analysis and related studies do show that increasing energy intake among severe COPD cases is difficult to accomplish, and if energy intake is not increased, weight and functionality will certainly not improve. Interventions should also be extended to prevention and early treatment of weight loss, that is, before patients are extremely wasted. This means expanding the target group to include COPD out-patients and primary care patients before they have become underweight, and putting more emphasis on dietary change than on medically-prescribed supplementation. Few studies have been published on the possibilities and effects of voluntary dietary change among out-patients. Diet is often part of the focus in self-help or self-management programs for COPD patients and some of these programs have been evaluated. However there have not yet been any well-controlled studies of the prevention of weight loss in COPD.

Nutritional modulation

Body compositional studies have shown that weight loss is accompanied by significant loss of fat-free mass and that it is specifically the loss of fat-free mass or other measures of muscle mass that are related to impaired skeletal muscle strength and exercise capacity [15, 16]. These studies have furthermore shown that muscle wasting may also occur in normal weight stable subjects. A recent study even suggested that muscle mass is a better predictor of survival than body weight [17]. Wasting of muscle mass is due to an impaired balance between protein synthesis (anabolism) and protein breakdown (catabolism). Besides nutritional abnormalities and physical inactivity, altered neuro-endocrine response and presence of a systemic inflammatory response may contribute to a negative protein balance in chronic diseases. From a therapeutic perspective it is important to know the relative contribution of these factors to altered protein synthesis and protein breakdown respectively. While increasing dietary intake can compensate elevated energy requirements and *vice versa*, uncontrolled protein breakdown cannot be overcome by only increasing protein synthesis and *vice versa*.

Several studies have investigated in COPD and other chronic wasting disorders the effects of pharmacological anabolic stimuli to promote protein synthesis including anabolic steroids, growth hormone and insulin-like growth factor. No studies have yet specifically investigated the ability to induce or enhance muscle weight gain by nutritional modulation of protein synthesis or protein breakdown rates. Optimising protein intake and essential amino acid intake may stimulate protein synthesis *per se*, but also enhance efficacy of anabolic drugs [20] as well as physiological stimuli such as resistance exercise [21]. Further studies are also indicated to investigate whether an anabolic response may be enhanced by modulation of cellular metabolism of the muscle by bio-active nutrients involved in muscle energy and substrate metabolism such as creatine, carnitine, antioxidants and amino acids. This exciting new area of research may shift the focus of nutritional therapy from merely supportive care to modulation of oxidative metabolism in order to induce or enhance the response to pulmonary rehabilitation.

References

1 Siafakas NM, Vermeire P, Pride NB, Paoletti P, Gibson J, Howard P, Yernault JC, Decramer M, Higenbottam T, Postma DS et al (1995) Optimal assessment and management of chronic obstructive pulmonary disease. ERS consensus statement. *Eur Respir J* 8: 1398–1420

2 American Thoracic Society, ATS statement (1995) Standards for the diagnosis and care of patients with chronic obstructive pulmonary disease. *Am J Respir Crit Care Med* 152: S77–S121

3 Petty TL (1976) Pulmonary rehabilitation. *J Maine Med Assoc* 67: 199–205

4 Fishman AP (1994) Pulmonary rehabilitation research. *Am J Respir Crit Care Med* 149: 825–833

5 Donner CF, Muir JF (1997) Selection criteria and programmes for pulmonary rehabilitation in COPD patients: ERS Task Force Position Paper. *Eur Respir J* 10: 744–757

6 American Thoracic Society (1999) Pulmonary rehabilitation: Official statement of the American Thoracic Society. *Am J Respir Crit Care Med* 159: 1666–1682

7 World Health Organisation (1980) International Classification of Impairments, disabilities and handicaps. World Health Organisation, Geneva

8 World Health Organisation (1999) *Towards a Common Language for Functioning and Disablement: ICIDH-2*. World Health Organization, Geneva: http://www.who.ch/icidh

9 Ries AL, Kaplan RM, Blumberg E (1991) Use of factor analysis to consolidate multiple outcome measures in chronic obstructive pulmonary disease. *J Clin Epidemiol* 44: 497–503

10 Wegner RE, Jorres RA, Kirsten DK, Magnussen H (1994) Factor analysis of exercise capacity, dyspnoea ratings and lung function in patients with severe COPD. *Eur Respir J* 7: 725–729

11 Mahler DA, Faryniarz K, Tomlinson D, Colice GL, Robins AG, Olmstead EM, O'Connor GT (1992) Impact of dyspnea and physiologic function on general health status in patients with chronic obstructive pulmonary disease. *Chest* 102: 395–401

12 Berry MJ, Rejeski WJ, Adair NE, Zaccaro D (1999) Exercise rehabilitation and chronic obstructive pulmonary disease stage. *Am J Respir Crit Care Med*; 160: 1248–1253

13 Maes S (1993) Chronische Ziekte [Chronic illnesses]. In: *Handboek Klinische Psychologie*, November. Houten, Bohn, Stafleu van Loghum, 1–35

14 Van den Broek AHS (1995) *Patient education and chronic obstructive pulmonary disease*. Thesis University of Leiden, Leiden

15 Ries AL, Kaplan RM, Limberg TM, Prewitt LM (1995) Effects of pulmonary rehabilitation on physiologic and psychosocial outcomes in patients with chronic obstructive pulmonary disease. *Ann Intern Med* 122 : 823–832

16 O'Donnell DE, McGuire M, Samis L, Webb KA (1995) The impact of exercise reconditioning on breathlessness in severe chronic airflow limitation. *Am J Respir Crit Care Med* 152 (6 Pt 1): 2005–2013

17 Goldstein RS, Gort EH, Stubbing D, Avendano MA, Guyatt GH (1994) Randomised controlled trial of respiratory rehabilitation. *Lancet* 344: 1394–1397
18 Strijbos JH, Postma DS, van Altena R, Gimeno F, Koeter GH (1996) A comparison between an outpatient hospital-based pulmonary rehabilitation program and a home-care pulmonary rehabilitation program in patients with COPD. A follow-up of 18 months. *Chest* 109: 366–372
19 Wijkstra PJ, van der Mark TW, Kraan J, van Altena R, Koeter GH, Postma DS (1996) Effects of home rehabilitation on physical performance in patients with chronic obstructive pulmonary disease (COPD). *Eur Respir J* 9: 104–110
20 Wedzicha JA, Bestall JC, Garrod R, Garnham R, Paul EA, Jones PW (1998) Randomized controlled trial of pulmonary rehabilitation in severe chronic obstructive pulmonary disease patients, stratified with the MRC dyspnoea scale. *Eur Respir J* 12: 363–369
21 Swerts PM, Kretzers LM, Terpstra Lindeman E, Verstappen FT, Wouters EF (1990) Exercise reconditioning in the rehabilitation of patients with chronic obstructive pulmonary disease: a short- and long-term analysis. *Arch Phys Med Rehabil* 71: 570–573
22 Cambach W, Chadwick-Straver RV, Wagenaar RC, van Keimpema AR, Kemper HC (1997) The effects of a community-based pulmonary rehabilitation programme on exercise tolerance and quality of life: a randomized controlled trial. *Eur Respir J* 10: 104–113
23 Bendstrup KE, Ingemann Jensen J, Holm S, Bengtsson B (1997) Out-patient rehabilitation improves activities of daily living, quality of life and exercise tolerance in chronic obstructive pulmonary disease. *Eur Respir J* 10: 2801–2806
24 Lacasse Y, Wong E, Guyatt GH, King D, Cook DJ, Goldstein RS (1996). Meta-analysis of respiratory rehabilitation in chronic obstructive pulmonary disease. *Lancet* 348: 1115–1119
25 Goldstein RS, Gort EH, Guyatt GH, Feeny D (1997) Economic analysis of respiratory rehabilitation. *Chest* 112: 370–379
26 Troosters T, Gosselink R, Decramer M (2000) Short- and long-term effects of outpatient rehabilitation in patients with chronic obstructive pulmonary disease: a randomized trial. *Am J Med* 109: 207–212
27 Griffiths TL, Burr ML, Campbell IA, Lewis Jenkins V, Mullins J, Shiels K, Turner Lawlor PJ, Payne N, Newcombe RG, Ionescu AA et al (2000) Results at 1 year of outpatient multidisciplinary pulmonary rehabilitation: a randomised controlled trial. *Lancet* 355: 362–368
28 Foglio K, Bianchi L, Ambrosino N (2001) Is it really useful to repeat outpatient pulmonary rehabilitation programs in patients with chronic airway obstruction? A 2-year controlled study. *Chest* 119: 1696–1704
29 Killian KJ, Leblanc P, Martin DH, Summers E, Jones NL, Campbell EJ (1992) Exercise capacity and ventilatory, circulatory, and symptom limitation in patients with airflow limitation. *Am Rev Respir Dis* 146: 935
30 Maltais F, Simard AA, Simard C, Jobin J, Desgagnes P, LeBlanc P (1996) Oxidative

capacity of the skeletal muscle and lactic acid kinetics during exercise in normal subjects and in patients with COPD. *Am J Respir Crit Care Med* 153: 288

31　Satta A, Migliori GB, Spanevello A, Neri M, Bottinelli R, Canepari M, Pellegrino MA, Reggiani C (1997) Fibre types in skeletal muscles of chronic obstructive pulmonary disease patients related to respiratory function and exercise tolerance. *Eur Respir J* 10: 2853

32　Pouw EM, Schols AM, van der Vusse GJ, Wouters EF (1998) Elevated inosine monophosphate levels in resting muscle of patients with stable chronic obstructive pulmonary disease. *Am J Respir Crit Care Med* 157: 453

33　Wuyam B, Payen JF, Levy P, Bensaidane H, Reutenauer H, Le Bas JF, Benabid AL (1992) Metabolism and aerobic capacity of skeletal muscle in chronic respiratory failure related to chronic obstructive pulmonary disease. *Eur Respir J* 5: 157

34　Levine S, Kaiser L, Leferovich J, Tikunov B (1997) Cellular adaptation in the diaphragm in chronic obstructive pulmonary disease. *N Engl J Med* 337: 1799

35　O'Donnell DE, Webb KA (1993) Exertional breathlessness in patients with chronic airflow limitation. The role of lung hyperinflation. *Am Rev Respir Dis* 148: 1351–1357

36　Casaburi R, Patessio A, Ioli F, Zanaboni S, Donner CF, Wasserman K (1991) Reductions in exercise lactic acidosis and ventilation as a result of exercise training in patients with obstructive lung disease. *Am Rev Respir Dis* 143: 9

37　Maltais F, LeBlanc P, Simard C, Jobin J, Berube C, Bruneau J, Carrier L, Belleau R (1996) Skeletal muscle adaptation to endurance training in patients with chronic obstructive pulmonary disease. *Am J Respir Crit Care Med* 154: 442

38　Engelen MP, Schols AM, Does JD, Deutz NE, Wouters EF (2000) Altered glutamate metabolism is associated with reduced muscle glutathione levels in patients with emphysema. *Am J Respir Crit Care Med* 161: 98–103

39　Sala E, Roca J, Marrades RM, Alonso J, Gonzalez De Suso JM, Moreno A, Barbera JA, Nadal J, de Jover L, Rodriguez-Roisin R, Wagner PD (1999) Effects of endurance training on skeletal muscle bioenergetics in chronic obstructive pulmonary disease. *Am J Respir Crit Care Med* 159: 1726–1734

40　American college of Sports Medicine (1990) The recommended quantity and quality of exercise for developing and maintaining cardiorespiratory and muscular fitness in healthy adults. *Med Sci Sports Exerc* 23: 265–274

41　Casaburi R (1993) Exercise training in chronic obstructive lung disease. In: Casaburi R, Petty TL (eds): *Principles and practice of pulmonary rehabilitation*. WB Saunders, Philadelphia, 204–224

42　Green R, Singh S, Williams J, Morgan MA (2001) Randomised controlled trial of four weeks *versus* seven weeks of pulmonary rehabilitation in chronic obstructive pulmonary disease. Thorax 56: 143–145

43　Coppoolse R, Schols AM, Baarends EM, Mostert R, Akkermans MA, Janssen PP, Wouters EF (1999) Interval *versus* continuous training in patients with severe COPD. *Eur Respir J* 14: 258–263

44　Rooyackers J, Dekhuijzen P, van Herwaarden C, Folgering H (1997) Ventilatory response to positive and negative work in patients with chronic obstructive pulmonary disease. *Respir Med* 91: 143–149

45 Rooyackers J (1996) *Pulmonary rehabilitation in patients with severe chronic obstructive pulmonary disease*. Thesis Katholieke Universiteit Nijmegen, Nijmegen, The Netherlands

46 Ries AL, Kaplan RM, Limberg TM, Prewitt LM (1995) Effects of pulmonary rehabilitation on physiologic and psychosocial outcomes in patients with chronic obstructive pulmonary disease. *Ann Intern Med* 122: 823–832

47 Punzal PA, Ries AL, Kaplan RM, Prewitt LM (1991) Maximum intensity exercise training in patients with chronic obstructive pulmonary disease. *Chest* 100: 618–623

48 Ries AL, Archibald CJ (1987) Endurance exercise training at maximal targets in patients with chronic obstructive pulmonary disease. *J Cardiopulm Rehab* 7: 594–601

49 Maltais F, LeBlanc P, Jobin J, Berube C, Bruneau J, Carrier L, Breton MJ, Falardeau G, Belleau R (1997) Intensity of training and physiologic adaptation in patients with chronic obstructive pulmonary disease. *Am J Respir Crit Care Med* 155: 555–561

50 Clark C, Cochrane L, Mackay E (1996) Low intensity peripheral muscle conditioning improves exercise tolerance and breathlessness in COPD. *Eur Respir J* 9: 2590–2596

51 Vallet G, Ahmaidi S, Serres I, Fabre C, Bourgouin D, Desplan J, Varray A, Prefaut C (1997) Comparison of two training programmes in chronic airway limitation patients: standardized *versus* individualized protocols. *Eur Respir J* 10: 114–122

52 Simpson K, Killian K, McCartney N, Stubbing DG, Jones NL (1992) Randomised controlled trial of weightlifting exercise in patients with chronic airflow obstruction. *Thorax* 47: 70–75

53 Clark CJ, Cochrane LM, Mackay E, Paton B (2000) Skeletal muscle strength and endurance in patients with mild COPD and the effects of weight training. *Eur Respir J* 15: 92–97

54 Bernard S, Whittom F, Leblanc P, Jobin J, Belleau R, Berube C, Carrier G, Maltais F (1999) Aerobic and strength training in patients with chronic obstructive pulmonary disease. *Am J Respir Crit Care Med* 159: 896–901

55 Couser JI, Maryinez FJ, Celli BR (1992) Respiratory response and ventilatory muscle recruitment during arm elevation in normal subjects. *Chest* 101: 336–340

56 Celli BR, Rassulo J, Make BJ (1986) Dyssynchroneous breathing during arm but not leg exercise in patients with chronic airflow obstruction. *N Engl J Med* 314: 1485–1490

57 Criner GJ, Celli BR (1988) Effect of unsupported arm exercise on ventilatory muscle recruitment in patients with severe airflow obstruction. *Am Rev Respir Dis* 138: 856–861

58 Dolmage TE, Maestro L, Avendano MA, Goldstein RS (1993) The ventilatory response to arm elevation of patients with chronic obstructive pulmonary disease. *Chest* 104: 1097–1100

59 Martinez FJ, Couser Jl, Celli BR (1991) Respiratory response to arm elevation in patients with chronic airflow obstruction. *Am Rev Respir Dis* 143: 476–480

60 Baarends EM, Schols AM, Slebos DJ, Mostert R, Janssen PP, Wouters EF (1995) Meta-

bolic and ventilatory response pattern to arm elevation in patients with COPD and healthy age-matched subjects. *Eur Respir J* 8: 1345–1351

61   Ries AL, Ellis B, Hawkins R (1988) Upper extremity exercise training in chronic obstructive pulmonary disease. *Chest* 93: 688–692

62   Marchand E, Decramer M (2000) Respiratory muscle function and drive in chronic obstructive pulmonary disease. *Clin Chest Med* 21: 679–692

63   Smith K, Cook D, Guyatt GH, Madhavan J, Oxman AD (1992) Respiratory muscle training in chronic airflow limitation: a meta-analysis. *Am J Respir Crit Care Med* 145: 533–539

64   Harver A, Mahler DA, Daubenspeck JA (1989) Targeted inspiratory muscle training improves respiratory muscle function and reduces dyspnea in patients with chronic obstructive pulmonary disease. *Ann Intern Med* 111: 117

65   Dekhuijzen PNR, Folgering THM, Van Herwaarden CLA (1991) Target-flow inspiratory muscle training during pulmonary rehabilitation in patients with COPD. *Chest* 99: 128

66   Wanke T, Formanek D, Lahrmann H, Brath H, Wild M, Wagner C, Zwick H (1994) Effects of combined inspiratory muscle and cycle ergometer training on exercise performance in patients with COPD. *Eur Respir J* 7: 2205

67   Benditt JO, Wood DE, McCool FD, Lewis S, Albert RK (1997) Changes in breathing and ventilatory muscle recruitment patterns induced by lung volume reduction surgery. *Am J Respir Crit Care Med* 155: 279

68   Larson JL, Covey MK, Wirtz SE, Berry JK, Alex CG, Langbein WE, Edwards L (1999) Cycle ergometer and inspiratory muscle training in chronic obstructive pulmonary disease. *Am J Respir Crit Care Med* 160: 500

69   Damoiseaux V (1984) *Patiëntenvoorlichting: een terreinverkenning*. [Patient education: an exploration]. *Symposiumbundel patiëntenvoorlichting* 1984. G.V.O. cahiers, University of Maastricht, Maastricht

70   Jones K, Tilford S, Robinson Y (1990) *Health education. Effectiveness and efficiency*. Chapman and Hall, India

71   Jensen PS (1983) Risk, protective factors, and supportive interventions in chronic airway obstruction. *Arch Gen Psychiatry* 40: 1203–1207

72   Atkins CJ, Kaplan RM, Timms RM, Reinsch S, Lofback K (1984) Behavioural exercise programs in the management of chronic obstructive pulmonary disease. *J Consult Clin Psychol* 52: 591–603

73   Howland J, Nelson EC, Barlow PB, McHugo G, Meier FA, Brent P, Laser-Wolston N, Parker HW (1986) Chronic obstructive airway disease. Impact of health education. *Chest* 90: 233–238

74   Tougaard L, Krone T, Sorknaes A, Ellegaard H (1992) Economic benefits of teaching patients with chronic obstructive pulmonary disease about their illness. *Lancet* 339: 1517–1520

75 Toshima MT, Kaplan RM, Ries AL (1990) Experimental evaluation of rehabilitation in chronic obstructive pulmonary disease: short-term effects on exercise endurance and health status. *Health Psychol* 9: 237–252
76 Kaptein AA, Dekker FW (2000) Psychosocial support. *Eur Respir Mon* 5: 58–69

New drugs for COPD based on advances in pathophysiology

Trevor T. Hansel[1], Rachel C. Tennant[1], Edward M. Erin[1], Andrew J. Tan[1], Peter J. Barnes[2]

[1]National Heart and Lung Institute (NHLI) Clinical Studies Unit, Royal Brompton Hospital, Fulham Road, London SW3 6HP, UK; [2]National Heart and Lung Institute (NHLI), Department of Thoracic Medicine, Imperial College School of Medicine, Dovehouse Street, London SW3 6LY, UK

The need to develop new drugs

There is a pressing need to develop new treatments for chronic obstructive pulmonary disease (COPD), as no currently available drug has been shown to reduce the relentless progression of this disease. Furthermore, recognition of the global importance and rising prevalence of COPD and the absence of effective therapies has now led to a concerted effort to develop new drugs for this disease [1, 2]. However, there have been disappointingly few therapeutic advances in the drug therapy of COPD, in contrast to the enormous advances made in asthma management that reflect a much better understanding of the underlying disease [3, 4].

Clinical pathophysiology

Rational therapy depends on understanding the underlying disease process and there have been recent advances in understanding the cellular and molecular mechanisms that may be involved. There is a particular need to develop drugs that suppress the underlying inflammatory, fibrotic and destructive processes that underlie this disease (Fig. 1). At separate anatomical sites different pathological events occur in COPD (Fig. 2), with distinct physiological sequelae, that result in varying clinical manifestations. (1) Chronic bronchitis comprises chronic inflammation of the central airways and results in mucus hypersecretion, and a chronic productive cough. (2) Obstructive bronchiolitis involves inflammation of the peripheral airways resulting in local airway wall fibrosis and remodelling and manifests as obstructive airways disease. (3) Emphysema has proteolytic destruction together with fibrosis and remodelling in the respiratory bronchioles and alveoli. Lung parenchymal destruction and remodelling causes various types of emphysema, and in advanced cases impaired gas exchange results in hypoxic respiratory failure. (4) Pulmonary vascular disease and *cor pulmonale* involve destruction of the pulmonary capillary bed and inflammation of pulmonary arterial vessels causing pulmonary arterial hypertension and right-sided heart failure. (5) Systemic disease is extra-pulmonary inflammatory disease in advanced COPD, causing cachexia and loss of fat-free mass (FFM), with respiratory and peripheral muscle weakness.

COPD involves a chronic inflammation in small airways and lung parenchyma, with the involvement of macrophages, neutrophils, and CD8+ T lymphocytes (Fig. 3). Following chronic exposure to oxidants in cigarette smoke and other inhaled noxious agents, patients with COPD have an amplified inflammatory response in the airways and lung parenchyma [5]. Epithelial cell injury and macrophage activation causes release of chemotactic factors that recruit neutrophils from the circulation. Macrophages and neutrophils then release proteases, with involvement of matrix metalloproteases (MMPs) and neutrophil elastase (NE) that break down connective tissue. Cytotoxic CD8+ T cells may also be involved in this inflammatory cascade. Over many years of injury, cycles of inflammation and repair occur that may result in resolution, but that can be associated with proteolysis, fibrosis, and both airway and parenchymal remodelling. The inflammation of COPD is quite different from that seen in asthma, indicating that different treatments are likely to be needed [6, 7]. The nature of the inflammatory infiltrate is broadly similar in large and small airways as well as within the alveoli and pulmonary artery wall.

The pathology of COPD

The pathology of COPD has been the subject of a number of recent reviews [4, 7–13] as well as being covered in textbooks of lung pathology [14]. In advanced COPD chronic bronchitis, obstructive bronchiolitis, emphysema, pulmonary vascular disease and systemic disease may all occur in the same patient (Figs. 1 and 2).

Chronic bronchitis

At post mortem the bronchi of a patient with chronic bronchitis contain an abundance of mucus, that overlies a dusky red mucous membrane [14]. Chronic bron-

Figure 1
Pathophysiology of COPD. Oxidants in cigarette smoke are the major cause of COPD, but a variety of other factors contribute. An amplified inflammatory response is a feature of COPD, and is associated with cycles of resolution, mucus production, fibrosis and proteolysis. The pathology of COPD involves five sites: the large central airways (chronic bronchitis), the small peripheral airways (obstructive bronchiolitis), the lung parenchyma (emphysema), the cardiovascular system (cor pulmonale) and respiratory and peripheral muscles (systemic disease).

New drugs for COPD based on advances in pathophysiology

Causative factors: cigarette smoke and biomass fuel oxidants, industrial pollution, motor exhaust emissions, mineral dusts and particulates

→ inflammation (further oxidants)

Airway remodelling: resolution | mucus production | fibrosis | proteolysis | vascular disease | systemic disease

Pathology: nil | chronic bronchitis | obstructive bronchiolitis | emphysema | cor pulmonale | systemic inflammation and muscle pathology

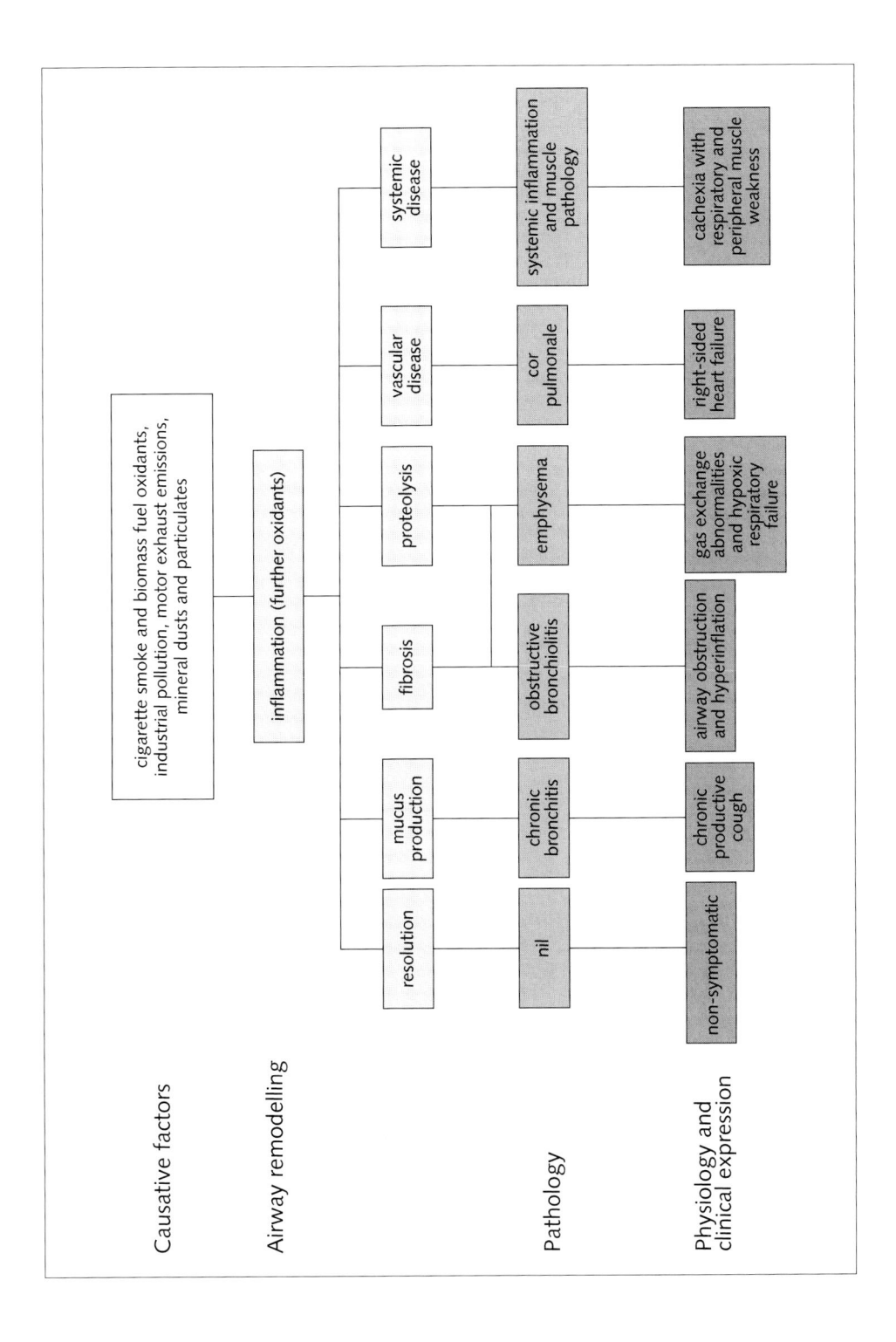

Physiology and clinical expression: non-symptomatic | chronic productive cough | airway obstruction and hyperinflation | gas exchange abnormalities and hypoxic respiratory failure | right-sided heart failure | cachexia with respiratory and peripheral muscle weakness

chitis affects mainly the intermediate sized bronchi with an internal diameter of 2–4 mm. Increased mucus is present on the surface of the lumen and occurs with goblet cell hyperplasia and enlarged submucosal glands. Enlargement of the submucosal bronchial glands is a major histological feature of chronic bronchitis [15, 16]. The Reid index describes the relative thickness of the submucosal mucus glands layer relative to that of the airway wall from base of epithelium to the inner cartilage surface. The normal index is 0.3, the submucosal glands occupying about a third of the airway wall thickness. Inflammation occurs in the mucosa as well as in the smooth muscle and submucosal glands.

Mucus hypersecretion may arise through activation of sensory nerve endings in the airways with reflex (local peptidergic and spinal cholinergic) increase in mucus secretion and direct stimulatory effects of enzymes such as neutrophil elastase and chymase. Recent studies suggest that epidermal growth factor (EGF) is a key mediator of mucus hyperplasia and mucus hypersecretion and may be the final common pathway that mediates the effects of many stimuli, including cigarette smoke, on mucus secretion [17]. With time there is hyperplasia of submucosal glands and proliferation of goblet cells under the influence of growth factors such as EGF. Chronic stimulation leads to up-regulation of mucin (MUC) genes. There are at least eight MUC genes now recognised in humans [18], but it is not yet certain which genes are over-expressed in chronic bronchitis. It was once thought that mucus hypersecretion played no role in airflow obstruction, but epidemiological data show that mucus hypersecretion is a risk factor for airflow obstruction and it is likely that viscous mucus may contribute to reduced airflow [19]. However, a 15-year study has found that GOLD Stage O with chronic bronchitis, does not identify subsequent airways obstruction [20].

Obstructive bronchiolitis

Obstructive bronchiolitis involves the small or peripheral airways, and is an inflammatory condition of the small airways. In obstructive bronchiolitis there is a collapsed lumen with increased mucus, unlike in asthma where the lumen is maintained. Within the epithelium Clara cells are replaced by goblet cells [21], and mucus appears in peripheral airways by the process of goblet cell metaplasia [22–23]. The replacement of the normal surfactant lining by mucus leads to an abnormally high surface tension and small airway instability, and predisposes to early airway closure during expiration [24]. Bronchial biopsies have demonstrated an infiltration with mononuclear cells; macrophages and cytotoxic (CD8+) T lymphocytes, rather than neutrophils, [3, 10, 25, 26]. In studies that examined the peribronchiolar inflammation of smokers whose lungs had been resected for localized tumor, those with chronic bronchitis and COPD had increased numbers of CD8+ cells [27, 28]. These inflammatory changes to small airways appear to be related to

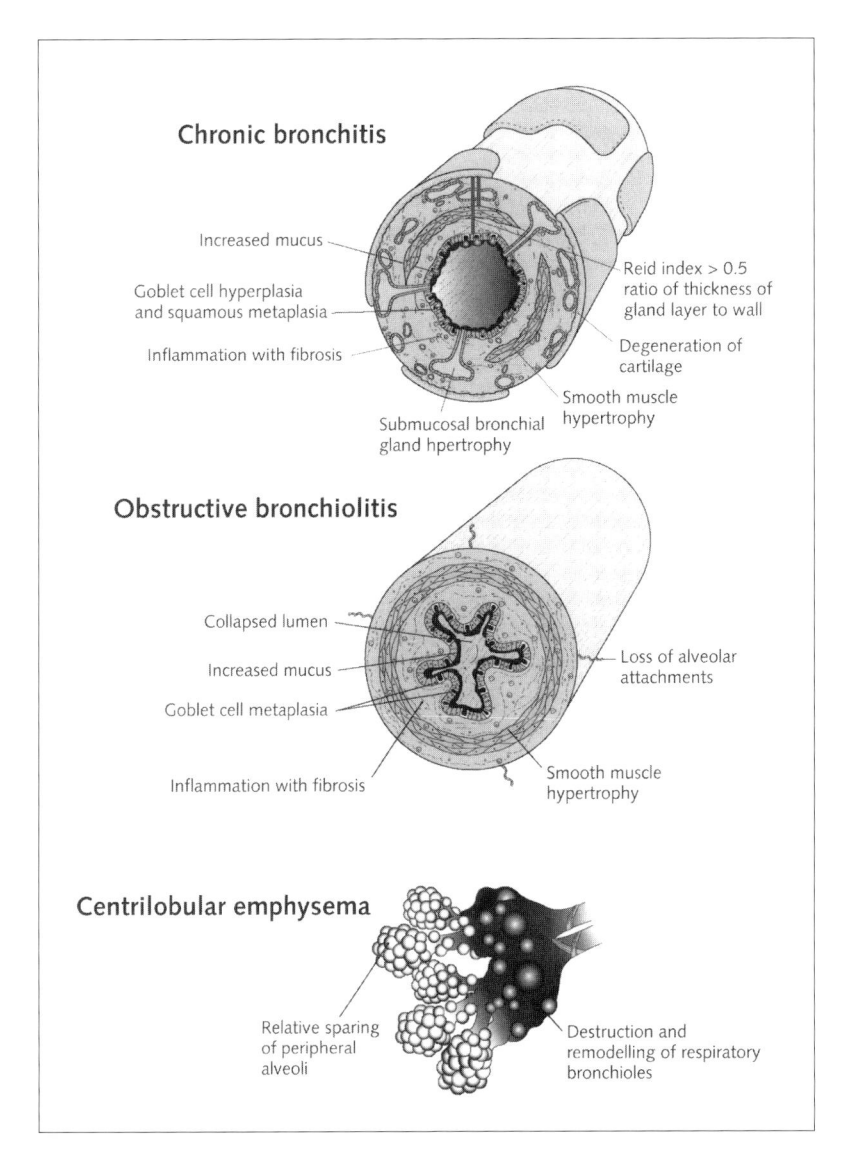

Figure 2
Pathological features of COPD. Chronic bronchitis involves goblet cell hyperplasia of the epithelium and submucosal bronchial gland hypertrophy, manifested clinically as mucus hypersecretion. Obstructive bronchiolitis is an inflammatory and fibrotic disease of the small airways that causes airway obstruction. Centilobular emphysema is characteristic of cigarette smoking and causes destruction and remodeling of the respiratory bronchioles and more central alveoli. Emphysema involves the destruction of the lung parenchyma and capillary bed and becomes manifest as respiratory failure due to failure of blood oxygenation.

clinical airflow obstruction in COPD, and this association with loss in FEV_1 appears to be stronger than that seen in the bronchi [26, 29–32]. Histologically, one of the most consistently observed early effects of cigarette smoke in patients with COPD is a marked increase in the number of macrophages in the bronchioli, and an associated respiratory bronchiolitis and alveolitis [33]. Increased numbers of macrophages can also be detected in bronchoalveolar lavage (BAL) [23, 34, 35].

There are increased numbers of fibroblasts and myofibroblasts, and increased extracellular matrix. The injury and repair process results in a structural remodeling of the airway wall, with increased collagen content and scar formation, that narrows the lumen and produces fixed airways obstruction [36]. The resultant stenotic narrowing of bronchioles has been convincingly demonstrated [37], and has been documented in terms of peripheral airways resistance [38]. Fibrosis in the small airways is characterised by the accumulation of mesenchymal cells (fibroblast and myofibroblasts) and extracellular connective tissue matrix.

Further structural changes in obstructive bronchiolitis include smooth muscle hypertrophy, mural edema, and an increased number of airways that are < 400 µm

Figure 3
Targets for future COPD therapy. The various targets for novel drugs to treat COPD are rationally based on understanding of the oxidant, inflammatory, fibrotic, proteolytic and regenerative processes that occur in the airways and lung parenchyma. Pathogenic pathways are shown in grey boxes, while potential therapy are in white boxes. Oxidants within cigarette smoke, as well as other irritants, activate epithelial cells and macrophages in the respiratory tract which release neutrophil chemotactic factors, including interleukin-8 (IL-8) and leukotriene B_4 (LTB_4). The IL-8 family causes chemotaxis of neutrophils via the cystein-X-cysteine receptor 2 (CXCR2), while monocyte chemotactic peptide-1 (MCP-1) binds to cysteine-cysteine receptor 2 (CCR2). Cytotoxic $CD8^+$ T cells may also be involved in the inflammatory cascade. Neutrophils and tissue macrophages are a potent source of oxidants ($O_2^{\cdot-}$) and proteases, the latter being normally inhibited by a panel of endogenous protease inhibitors. Proteases include neutrophil elastase (NE), cathepsins and various matrix metalloproteinases (MMPs). Protease inhibitors include α_1-antitrypsin (α_1-AT), secretory leukoprotease inhibitor (SLPI) and tissue inhibitor of matrix metalloproteinases (TIMP). Fibroblasts are a prominent feature in the pathology of bronchiolitis and emphysema, and regenerative healing processes, as well as tissue remodeling, occur in COPD. Vascular endothelial cells, pneumocytes and mast cells may also contribute to the pathogenesis of COPD.
CACC, calcium activated chloride channel; EGF, epidermal growth factor; FGF, fibroblast growth factors; GF, hepatocyte growth factor; iNOS, inducible nitric oxide synthase; ICAM-1, intercellular adhesion molecule-1; IL-10, interleukin-10; MAPKs, mitogen-activated protein kinases; NF-κB, nuclear factor kappa B; PPAR-γ, peroxisome proliferator-activated receptor-γ; PDE4, phosphodiesterase type 4; PI-3Kγ, phosphoinositide-3 kinase γ; PAR, protease activated receptor; TGF-β, transforming growth factor-β; TNF-α, tumour necrosis factor-α.

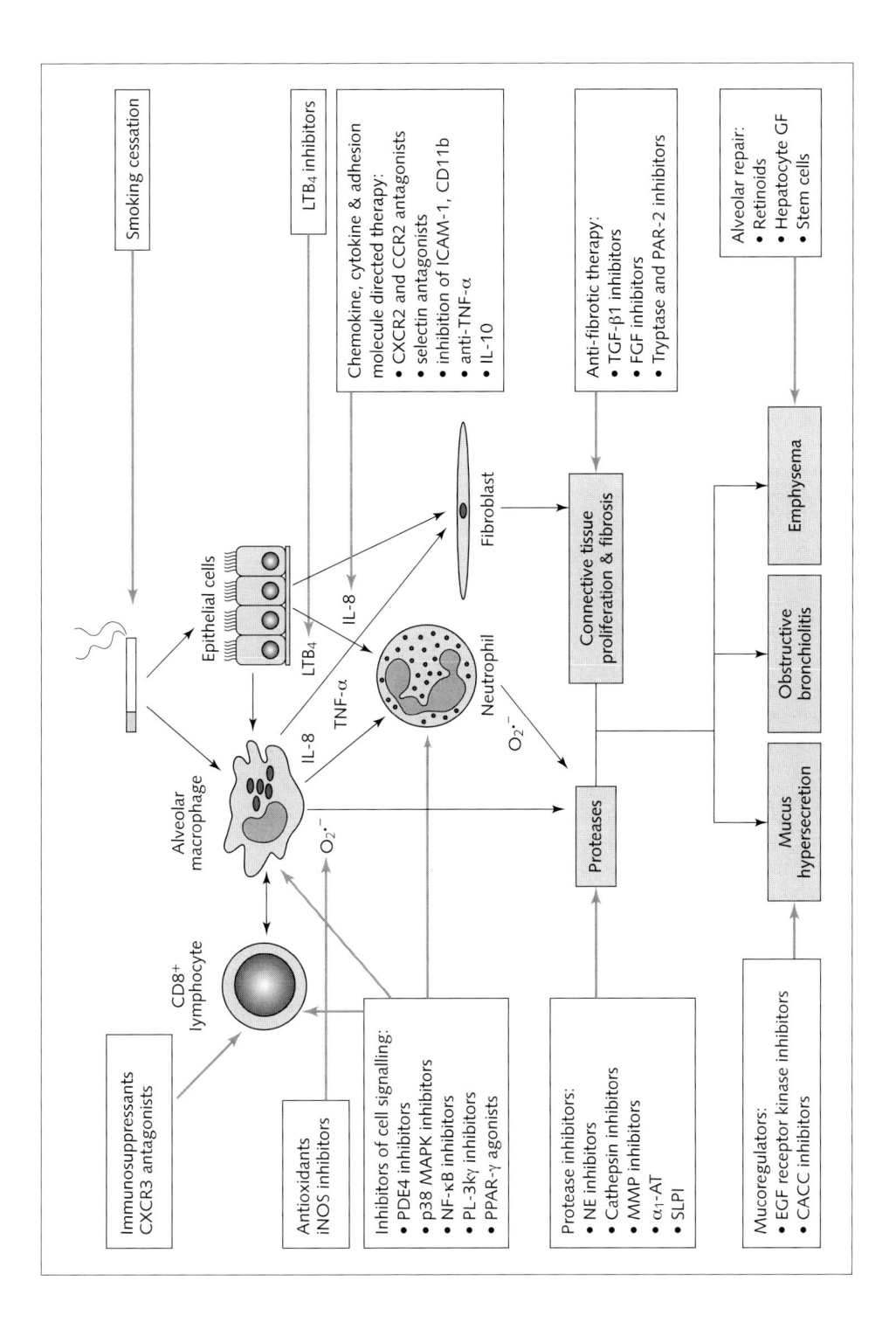

New drugs for COPD based on advances in pathophysiology

in diameter [23, 34, 39, 40]. Smooth muscle hypertrophy is most striking in small bronchi and bronchioli [27, 41]. In the small airways the inflammation moves into the interstitium to destroy the alveolar attachments that provide parenchymal support to the bronchiole. A loss of alveolar attachments to the bronchiole perimeter contributes to the loss of elastic recoil and favours increased tortuosity and early closure of bronchioles during expiration in patients with COPD [42–45].

There is debate about whether the airflow obstruction in COPD is primarily due to obstruction of the lumen of small airways as a result of bronchiolitis and fibrosis [46], or whether it is due to loss of elasticity and closure of small airways as a result of parenchymal destruction and loss of alveolar attachments. Loss of lung elasticity may occur in COPD even in the absence of emphysema and it is likely that this may be an important contributor to airway obstruction in all patients with COPD. The extent to which peripheral airway obstruction as a result of fibrosis and lumenal obstruction with mucus contributes to airflow limitation is likely to vary from patient to patient, but it is likely that both mechanisms contribute to a variable extent in most patients as cigarette smoking causes both abnormalities.

Emphysema

Emphysema is defined by permanent, destructive enlargement of the airspaces distal to the terminal bronchioli, affecting the respiratory bronchioles and sometimes the alveoli (Fig. 2). The enlargement is best not called a dilatation, since this implies a reversible bronchodilation as occurs in asthma. The mechanism of this process is poorly understood, but it is thought to be an inflammatory condition of the lung parenchyma mediated by T lymphocytes, neutrophils, and alveolar macrophages. Inflammation is associated with the release of excessive amounts of proteolytic enzymes such as neutrophil elastase (NE) and matrix metalloproteases (MMPs). Inflammation and proteolysis is accompanied by destruction of lung parenchyma, fibrosis and remodelling.

Centrilobular emphysema is the most common cause of cigarette smoking-induced emphysema in COPD. It involves the dilatation and destruction of the respiratory bronchioles [47, 48]. Centrilobular emphysema occurs more frequently in the upper lung fields in mild disease. In more advanced disease, lesions are more diffuse and also involve destruction of the capillary bed. In post mortem necropsy material, holes are visible in lung sections in patients with severe emphysema. The early changes of emphysema may include subtle disruption to elastic fibers with an accompanying loss of elastic recoil, bronchiolar and alveolar distortion. It has been postulated that fenestrae (pores of Kohn) may enlarge to develop microscopic emphysema [49, 50]. Lymphocytes have been demonstrated to form a significant component of the alveolar wall inflammatory infiltrate in COPD [51, 52]. The greater the number of T lymphocytes, the less alveolar tissue is present [53]. Neutrophil elastase is a powerful proteolytic enzyme, and neutrophils have also been implicated in lung damage.

The destructive process is accompanied by a net increase in the mass of collagen with alveolar wall fibrosis even in emphysematous lungs [54]. Emphysema is due to enzymatic destruction of the alveolar walls and destruction of alveolar attachments to the bronchioles. This results in the loss of driving pressure and airway narrowing with a consequent reduction in FEV_1. The fundamental defect is ventilation/perfusion (Va/Q) imbalance that becomes reflected in abnormalities in gas transfer, that eventually manifests as respiratory failure.

Pulmonary vascular disease

Pulmonary vascular changes in COPD begin early in the course of COPD as intimal thickening, followed by smooth muscle hypertrophy and inflammatory infiltration. This may be followed by pulmonary hypertension and destruction of the capillary bed. Endothelial dysfunction of the pulmonary arteries may be caused directly by cigarette smoke products [55] or indirectly by inflammatory mediators [56]. Endothelial dysfunction may then initiate the sequence of events that results in structural changes [57–59] possibly mediated through endothelial derived relaxing factors.

Early in the natural history of COPD, thickening of the intima begins in the walls of the muscular pulmonary arteries. This intimal hyperplasia is believed to occur when lung function is reasonably well maintained, and when pulmonary vascular pressures are normal at rest [60]. An increase in vascular smooth muscle occurs, although only a moderate degree of muscular hypertrophy has been reported [40], with extension of vascular smooth muscle to vessels that normally lack muscle. Infiltration of the vessel wall by inflammatory cells including macrophages and $CD8^+$ T cells occurs [53, 56]. Increasing amounts of smooth muscle, proteoglycans and collagen further thicken the vessel wall, and fibrosis may obliterate some vessels.

Hypoxic pulmonary vasoconstriction may contribute to pulmonary hypertension, while advanced COPD is associated with emphysematous destruction of the pulmonary capillary bed [58]. The mechanisms of vascular remodelling are not yet understood and it is likely that several growth factors, including vascular endothelial growth factor and fibroblast growth factor may be involved. Endothelin-1 (ET-1) is strongly expressed in pulmonary vascular endothelium of patients with pulmonary hypertension secondary to chronic hypoxia [61] and urinary ET-1 excretion is increased in patients with COPD [62] and sputum ET-1 levels are increased in exacerbations [63]. ET-1, acting mainly *via* ET_A receptors, induces fibrosis and hyperplasia of pulmonary vascular smooth muscle, implying a role in the pulmonary hypertension secondary to COPD.

Although pulmonary hypertension and *cor pulmonale* are common sequelae of COPD, the precise mechanisms of increased vascular resistance are unclear [58, 64]. Structural changes in the pulmonary arteries are correlated with an increase in pulmonary vascular pressure (pulmonary hypertension) that develops first with exercise and then at rest. Chronic hypoxia results in widespread pulmonary vasoconstriction, excerbating pulmonary hypertension and causing right heart failure (cor pulmonale).

Systemic extrapulmonary effects

Systemic features of COPD include disturbances in metabolism with cachexia, as well as increased respiratory and skeletal muscle fatigue with wasting [65]. Particularly patients with predominant emphysema may develop profound weight loss, and this is a predictor of increased mortality that is independent of poor lung function [66]. Weight loss in COPD has been associated with increased levels of TNF-α and soluble TNF-α receptors [67–70]. The skeletal muscle weakness may exacerbate dyspnoea, and skeletal and respiratory muscle training is an important aspect of pulmonary rehabilitation. Improved nutrition and a short course of anabolic steroids have been shown to improve lung function in patients with COPD [71].

Anti-inflammatory strategies

There are multiple cells and mediators involved in the immunopathology of COPD (Figs. 3 and 4), and specialised anti-inflammatory strategies are required to deal with this particular type of inflammation.

Phosphodiesterase-4 inhibitors

PDE4 is the predominant PDE expressed in neutrophils, CD8+ cells and macrophages [72] (Fig. 5), suggesting that PDE4 inhibitors would be effective in controlling inflammation in COPD [73, 74]. Selective PDE4 inhibitors, such as cilomilast and roflumilast, are active in animal models of neutrophil inflammation [75, 76]. Cilomilast had promising beneficial clinical effects in a six-week study in patients with moderate to severe COPD [77], and larger studies are currently underway. Roflumilast appears to be well tolerated at doses that significantly inhibit TNF-α release from peripheral blood monocytes [78]. PDE4 inhibitors have been limited by side-effects, particularly nausea and other gastrointestinal effects, but it might be possible to develop more selective inhibitors in the future which are less likely to be dose-limited by adverse effects [79].

Several steps may be possible to overcome the limitation of the adverse event of nausea and vomiting. It now seems likely that vomiting is due to inhibition of a particular subtype of PDE4. At least four human PDE4 genes have been identified and each has several splice variants [80]. This raises the possibility that subtype-selective inhibitors may be developed that may preserve the anti-inflammatory effect, while having less propensity to side-effects. PDE4D appears to be of particular importance in nausea and vomiting and is expressed in the chemosensitive trigger zone in the brain stem [81] and in mice deletion of the gene for PDE4D prevents a behavioural equivalent of emesis [82]. This isoenzyme appears to be less important in anti-inflammatory effects and targeted gene disruption studies in mice indicate that PDE4B is more important than PDE4D in inflammatory cells [83]. PDE4B selective inhibitors may therefore have a greater therapeutic ratio and theoretically might be effective anti-inflammatory drugs. Cilomilast is the PDE4 inhibitor that has been most fully tested in clinical studies, particularly in COPD, but this drug is selective for PDE4D and therefore has a propensity to cause emesis. Roflumilast, which is a non-selective for PDE4 isoenzymes, looks more promising, as it has a more favourable therapeutic ratio [84]. Several other potent PDE4 inhibitors with a more favourable therapeutic ratio are now in clinical development for COPD.

Anti-oxidants

There is considerable evidence that oxidative stress is increased in patients with COPD and that reactive oxygen species (ROS) contribute to the pathophysiology of COPD, particularly during exacerbations [85–90]. Each puff of cigarette smoke contains of the order of 10^{17} ROS molecules, and ROS are also produced endogenously by activated inflammatory cells, including neutrophils and macrophages. A range of ROS cause a spectrum of effects in COPD. Increased levels of hydrogen peroxide (H_2O_2) are present in expired condensates from patients with COPD, particularly during exacerbations [91]. There is an increase in concentrations of ethane in exhaled air, this being a product of lipid peroxidation [90]. Oxidative stress leads to the formation of isoprostanes by direct oxidation of arachidonic acid, and 8-isoprostane is found in exhaled breath condensate of COPD patients at elevated levels that are related to disease severity [89, 92]. Isoprostanes have several effects on airway function, including bronchoconstriction, increased plasma leakage and mucus hypersecretion. Superoxide anions ($O_2^·$) rapidly combine with nitric oxide (NO) to form the potent radical peroxynitrite ($ONOO^-$), which itself generates $OH^·$. Peroxynitrite reacts with tyrosine residues within certain proteins to form 3-nitrotyrosines which may be detected immunologically, with increased reactivity in sputum macrophages from patients with COPD [87].

There is evidence for a reduction in antioxidant defenses in patients with COPD, which may further enhance oxidative stress [85, 88]. ROS are normally counteracted by endogenous (glutathione, uric acid, bilirubin) and exogenous (dietary vitamin C and E) antioxidants. Oxidants may contribute to the pathophysiology of COPD by induction of serum protease inhibitors, potentiation of elastase activity, direct activation of MMPs, and inactivation of anti-proteases such as α_1-antitrypsin (α_1-AT) and secretory leukoprotease inhibitor (SLPI). Oxidants thus both directly increasing proteolytic activity and decrease the antiprotease shield. Oxidants are potent mucus secretagogues, and activate the transcription factor nuclear factor-κB (NF-κB) which orchestrates the transcription of many inflammatory genes, including IL-8, TNF-α, inducible NO synthase (iNOS) and inducible cyclo-oxygenase (COX-2). Hydrogen peroxide directly constricts airway smooth muscle *in vitro*, while hydroxyl radicals (OH·) potently induce plasma exudation in airways.

This large body of evidence suggests that antioxidants may be of use in the therapy of COPD. N-acetyl cysteine (NAC) provides cysteine for enhanced production of the antioxidant glutathione (GSH) and has antioxidant effects *in vitro* and *in vivo*. Recent systematic reviews of studies with oral NAC in COPD suggest a small but significant reductions in exacerbations [93, 94]. More effective antioxidants, including stable glutathione compounds, analogues of superoxide dismutase and selenium-based drugs, are now in development for clinical use [86, 95].

Resveratrol is a phenolic component of red wine that has anti-inflammatory and antioxidant properties. It has a marked inhibitory effect on cytokine release from alveolar macrophages from COPD patients that show little or no response to corti-

Figure 4
Therapy directed against adhesion molecules, chemokines and cytokines. This figure represents the passage of neutrophils and monocytes from blood to lung tissue, with potential therapeutics shown in white boxes. Endothelial cell adhesion molecules are induced in response to inflammatory stimuli, and include P- and E-selectin, intercellular adhesion molecule-1 and -2 (ICAM-1 and -2). Neutrophils and monocytes express a variety of adhesion molecules including L-selectin, sialomucins and αMβ2 (CD11b, Mac-1), which bind to endothelial cells through adhesion molecule interactions. Interleukin-8 (IL-8) is a member of the CXC family of chemokines, and binds to CXCR1 as well as CXCR2 on neutrophils. Monocyte chemotactic peptide-1 (MCP-1) is a CC chemokine that binds to CCR2 on monocytes. Tumour necrosis factor-α (TNF-α) is a pro-inflammatory cytokine whose production is influenced by phosphodiesterase type 4 (PDE4) activity, and whose release from cells is governed by TNF-α converting enzyme (TACE). TNF-α is present as a homotrimer that binds to membrane-bound TNF receptors (TNFR) on a variety of different cells. Interleukin-10 (IL-10) is an endogenous anti-inflammatory cytokine that decreases the expression of chemokines such as interleukin-8 (IL-8) and monocyte chemotactic protein (MCP). In addition, subcutaneous IL-10 decreases levels of TNF-α and matrix metalloproteinases (MMPs), while increasing levels of tissue inhibitors of MMPs (TIMPs).

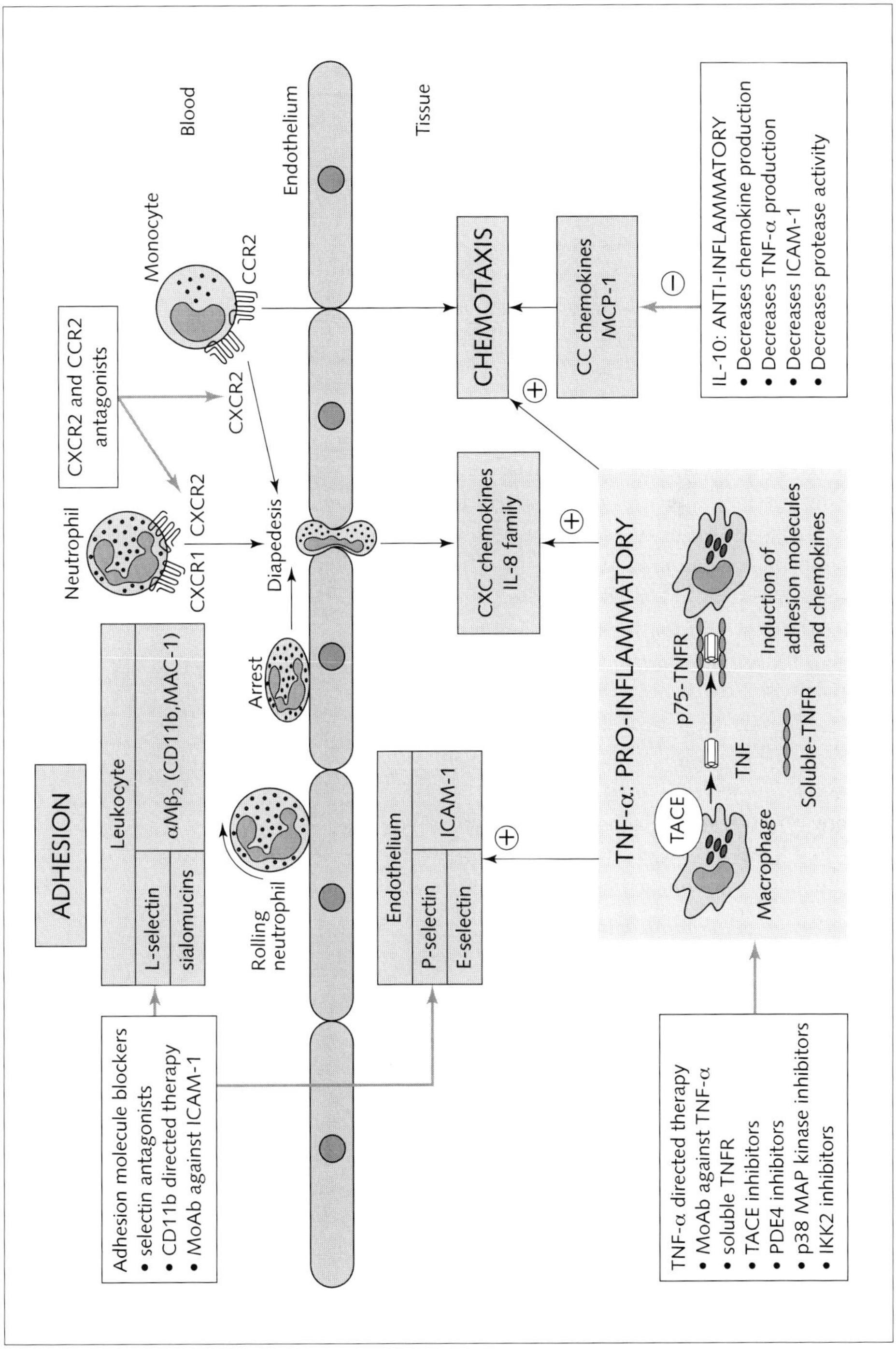

Blood
Endothelium
Tissue
ADHESION
Leukocyte
L-selectin
sialomucins
αMβ2 (CD11b,MAC-1)
Endothelium
P-selectin
E-selectin
ICAM-1
Adhesion molecule blockers
• selectin antagonists
• CD11b directed therapy
• MoAb against ICAM-1
Rolling neutrophil
Arrest
Diapedesis
Neutrophil
CXCR1
CXCR2
CXCR2 and CCR2 antagonists
CXCR2
Monocyte
CCR2
CXC chemokines
IL-8 family
CHEMOTAXIS
CC chemokines
MCP-1
IL-10: ANTI-INFLAMMATORY
• Decreases chemokine production
• Decreases TNF-α production
• Decreases ICAM-1
• Decreases protease activity
TNF-α: PRO-INFLAMMATORY
TACE
TNF
p75-TNFR
Macrophage
Soluble-TNFR
Induction of adhesion molecules and chemokines
TNF-α directed therapy
• MoAb against TNF-α
• soluble TNFR
• TACE inhibitors
• PDE4 inhibitors
• p38 MAP kinase inhibitors
• IKK2 inhibitors

costeroids [96]. The molecular mechanism of this action is currently unknown, but identification of the cellular target for resveratrol may lead to the development of a novel class of anti-inflammatory compounds. Resveratrol itself has a very low oral bioavailability so related drugs will need to be developed.

Oxidative stress and increased nitric oxide release from activity of inducible nitric oxide synthase (iNOS) may result in the formation of peroxynitrite; this is a potent radical that nitrates proteins and alters their function. 3-Nitrotyrosine may indicate peroxynitrite formation and is markedly increased in sputum macrophages of patients with COPD [87]. Selective inhibitors of iNOS are now in development [97] and one of these L-N^6-(1-imminoethyl)lysine (L-NIL) gives a profound and long-lasting reduction in the concentrations of nitric oxide in exhaled breath [98].

Leukotriene inhibitors

Leukotriene B$_4$ (LTB$_4$) is a potent chemoattractant of neutrophils and is increased in the sputum and exhaled breath of patients with COPD [99]. It is probably derived from alveolar macrophages as well as neutrophils and may be synergistic with IL-8. Alveolar macrophages from patients with α_1-antitrypsin deficiency secrete greater amounts of LTB$_4$ [100]. Two subtypes of receptor for LTB$_4$ have been described; BLT$_1$ receptors are mainly expressed on granulocytes and monocytes, whereas BLT$_2$ receptors are expressed on T lymphocytes [101]. BLT$_1$ antagonists, such as LY29311, have now been developed for the treatment of neutrophilic inflammation [102]. LY293111 and another antagonist SB225002 inhibit the neutrophil chemotactic activity of sputum from COPD patients, indicating the potential clinical value of such drugs [103, 104]. Several selective BLT$_1$ antagonists are now in development. LTB$_4$ is synthesised by 5'-lipoxygenase (5-LO), of which there are several

Figure 5
Inhibitors of cell signaling. Phosphodieterase type 4 (PDE4) catalyzes the inactivation of cyclic adenosine monophosphate (cAMP) to AMP, cAMP being an active second messenger within the cell. Multiple pathways mediate kinase transcription factor (TF) activity in relation to inflammatory gene expression, including a system acting on nuclear factor κB (NF-κB). Inhibitor of NF-κB (I-κB) kinases (IKK-2) and p38 mitogen-activated protein kinases (MAPKs) are involved. Inflammatory gene activation causes synthesis of cytokines, chemokines, adhesion molecules and proteases (for abbreviations see footnote to Fig. 3). Peroxisome proliferator-activated receptors (PPARs) are a family of hormone receptors that belong to the steroid superfamily, and PPAR-γ has anti-inflammatory activity. Phosphoinositide-3 (PI-3) kinase causes activation of cells through generation of phosphoinositide second messengers.

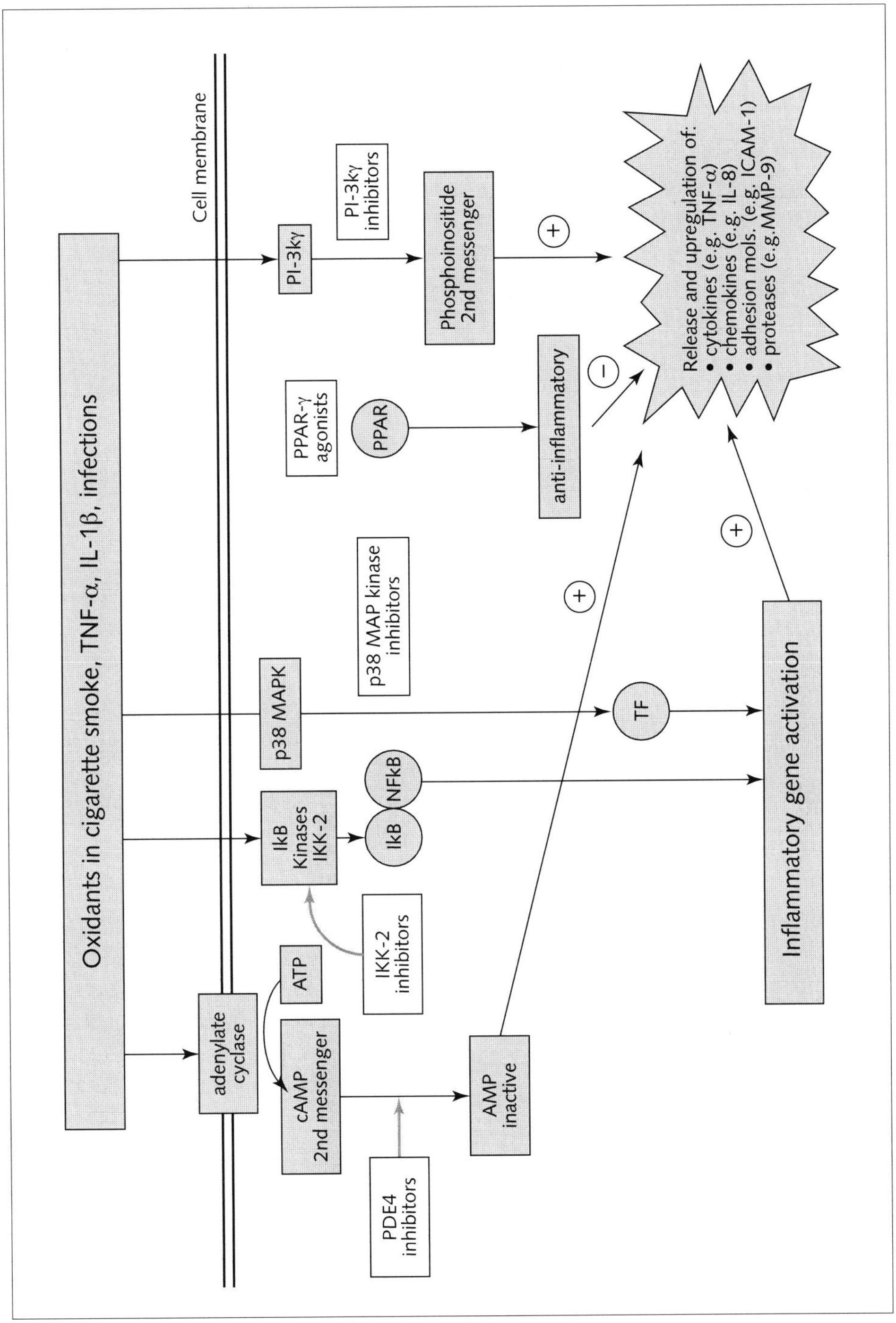
Oxidants in cigarette smoke, TNF-α, IL-1β, infections
Cell membrane
adenylate cyclase
ATP
cAMP 2nd messenger
PDE4 inhibitors
AMP inactive
IkB Kinases IKK-2
IKK-2 inhibitors
IkB
NFkB
p38 MAPK
p38 MAP kinase inhibitors
TF
Inflammatory gene activation
PPAR-γ agonists
PPAR
anti-inflammatory
PI-3kγ
PI-3kγ inhibitors
Phosphoinositide 2nd messenger
Release and upregulation of:
• cytokines (e.g. TNF-α)
• chemokines (e.g. IL-8)
• adhesion mols. (e.g. ICAM-1)
• proteases (e.g.MMP-9)

inhibitors, although there have been problems in clinical development of drugs in this class because of side-effects. A recent pilot study in COPD patients with a 5'-lipoxygenase inhibitor BAYx1005 showed only a modest reduction in sputum LTB_4 concentrations but no effect on neutrophil activation markers [105].

Adhesion molecule blockers

Recruitment of neutrophils, monocytes and cytotoxic T cells into the lungs and respiratory tract is dependent on adhesion molecules expressed by these cells and on endothelial cells in the pulmonary and bronchial circulations (Fig. 4). Several adhesion molecules can now be inhibited pharmacologically. For example, E-selectin on endothelial cells interacts with sialyl-Lewis^x on neutrophils. A mimic of sialyl-Lewis^x, TBC1269, blocks selectins and inhibits granulocyte adhesion, with preferential effects on neutrophils [106]. However, there are concerns about this therapeutic approach for a chronic disease, as an impaired neutrophilic response may increase the susceptibility to infection. The expression of Mac-1 (CD11b/CD18) is increased on neutrophils of patients with COPD, suggesting that targeting this adhesion molecule, which is also expressed on monocytes and macrophages, might be beneficial [107].

Chemokine inhibitors

Several chemokines are involved in neutrophil chemotaxis, these being mainly chemokines of the CXC family, and interleukin (IL)-8 receptor antagonists are of potential therapeutic benefit in COPD [108] (Fig. 4). IL-8 levels are markedly elevated in the sputum of patients with COPD and are correlated with disease severity [109, 110], and is also found in increased amounts in BAL fluid [111]. Blocking antibodies to IL-8 and related chemokines inhibit certain types of neutrophilic inflammation in experimental animals [112], and reduce the chemotactic response of neutrophils to sputum from COPD patients [99, 104, 113]. A human monoclonal antibody to IL-8 blocks the chemotactic response of neutrophils to IL-8 and is effective in animal models of neutrophilic inflammation [112]. This antibody is now in clinical trials for COPD, but it may be less effective than drugs that block the common receptor for other members of the CXC chemokine family. IL-8 signals through 2 receptors: a low affinity CXCR1 specific for IL-8 that is involved in neutrophil activation, and a high affinity CXCR2 that is activated by a range of CXC chemokines including IL-8, growth related oncogene (GRO-α,-β, -γ), and epithelial-derived neutrophil activating peptide, ENA-78 [114]. Other CXC chemokines, such as growth related oncoprotein-α (GRO-α), are also elevated in COPD [115] and therefore a CXCR2 antagonist is likely to be more useful than a CXCR1 antagonist, particularly as CXCR2 are also expressed on monocytes. Indeed, inhibition of monocyte chemotaxis may prevent the marked increase in macrophages found in the lungs of patients with COPD that may drive the inflammatory process. Small molecule inhibitors of CXCR2, such as SB225002, have now been developed and are entering clinical trials [116, 117].

The CC chemokine macrophage chemotactic peptide-1, MCP-1, is increased in BAL of COPD patients [118] and in macrophages and epithelial cells [119]. MCP-1 is a potent chemoattractant for monocytes, and acts *via* CCR2. GRO-α (growth related oncogene-α) is elevated in COPD [115] and is chemotactic for monocytes as well as neutrophils and may therefore contribute to the increased numbers of macrophages that are derived from blood monocytes in COPD.

Chemokine receptors are also important for the recruitment of CD8+ T cells which predominate in COPD airways and lungs and might contribute to the development of emphysema. CD8+ cells show increased expression of CXCR3 and there is up-regulation of CXCR3 ligands, such as CXCL10 (IP-10), in peripheral airways of COPD patients [120]. This suggests that CXCR3 antagonists might be useful.

Tumour necrosis factor-α (TNF-α) inhibitors

TNF-α levels are raised in the sputum of COPD patients [109], especially during exacerbations [121]. It is thought that TNF-α augments inflammation and induces IL-8 and other chemokines in airway cells and skeletal muscle *via* activation of the transcription factor NF-κB [122] (Fig. 4). The severe wasting in some patients with advanced COPD might be due skeletal muscle apoptosis, resulting from increased circulating TNF-α. COPD patients with cachexia have increased release of TNF-α from circulating leukocytes [67], and soluble TNF receptors are increased in sputum [123]. Humanised monoclonal TNF antibody (infliximab) and soluble TNF receptors (etanercept) that are effective in other chronic inflammatory diseases, such as rheumatoid arthritis and inflammatory bowel disease, should also be effective in COPD, particularly in patients who have systemic symptoms [124, 125]. Trials of anti-TNF therapies in patients with systemic features of COPD are currently underway. TNF-α converting enzyme (TACE), which is required for the release of soluble TNF-α, may be a more attractive target as it is possible to discover small molecule TACE inhibitors, some of which are also matrix metalloproteinase inhibitors [126, 127]. General anti-inflammatory drugs such as phosphodiesterase inhibitors and p38 mitogen-activated protein (MAP) kinase inhibitors [128] also potently inhibit TNF-α expression.

Interleukin-10

IL-10 is a cytokine with a wide spectrum of anti-inflammatory actions (Fig. 4). It inhibits the secretion of TNF-α and IL-8 from macrophages, and tips the balance in favour of antiproteases by decreasing the expression of matrix metalloproteinases, while increasing the expression of endogenous tissue inhibitors of matrix metalloproteinases (TIMP). IL-10 concentrations are reduced in induced sputum from patients with COPD, so that this may be a mechanism for increasing lung inflammation [129]. IL-10 is currently in clinical trials for other chronic inflammatory diseases (inflammatory bowel disease, rheumatoid arthritis and psoriasis), including patients with steroid resistance, but IL-10 may cause haematological side effects [130]. Treatment with daily injections of IL-10 over several weeks has been well tolerated. IL-10 may have therapeutic potential in COPD, especially if a selective activator of IL-10 receptors or unique signal transduction pathways can be developed in the future.

NF-κB inhibitors

NF-κB regulates the expression of IL-8 and other chemokines, TNF-α and other inflammatory cytokines, and some matrix metalloproteinases (Fig. 5). NF-κB is activated in macrophages and epithelial cells of COPD patients, particularly during exacerbations [131, 132]. There are several possible approaches to inhibition of NF-κB, including gene transfer of the inhibitor of NF-κB (IκB), a search for inhibitors of IκB kinases (IKK), NF-κB-inducing kinase (NIK) and IκB ubiquitin ligase, which regulate the activity of NF-κB, and the development of drugs that inhibit the degradation of IκB [133]. The most promising approach may be the inhibition of IKK-2 by small molecule inhibitors, several of which are now in development [134]. A small molecule IKK-2 inhibitor suppresses the release of inflammatory cytokines and chemokines from alveolar macrophages and might be effective in COPD when alveolar macrophages appear to be resistant to the anti-inflammatory actions of corticosteroids [135]. One concern about long-term inhibition of NF-κB is that effective inhibitors may result in immune suppression and impair host defences, since mice which lack NF-κB genes succumb to septicaemia. However, there are alternative pathways of NF-κB activation that might be more important in inflammatory disease [136].

p38 MAP kinase inhibitors

Mitogen-activated protein (MAP) kinases play a key role in chronic inflammation and several complex enzyme cascades have now been defined [137] (Fig. 5). One of these, the p38 MAP kinase pathway, is activated by cellular stress and regulates the expression of inflammatory cytokines, including IL-8, TNF-α and MMPs [138, 139]. Non-peptide inhibitors of p38 MAP kinase, such as SB 203580, SB 239063 and RWJ 67657, have now been developed and these drugs have a broad range of anti-inflammatory effects [128]. SB 239063 reduces neutrophil infiltration after inhaled endotoxin and the concentrations of IL-6 and MMP-9 in bronchoalveolar lavage fluid of rats, indicating its potential as an anti-inflammatory agent in COPD [140]. It is likely that such a broad spectrum anti-inflammatory drug will have some toxicity, but inhalation may be a feasible therapeutic approach.

Phosphoinositide 3-kinase inhibitors

PI-3Ks are a family of enzymes that lead to the generation of lipid second messengers that regulate a number of cellular events. A particular isoform, PI-3Kγ, is involved in neutrophil recruitment and activation. Knock-out of the PI-3Kγ gene results in inhibition of neutrophil migration and activation, as well as impaired T lymphocyte and macrophage function [141]. This suggests that selective PI-3Kγ inhibitors may have relevant anti-inflammatory activity in COPD and small molecule inhibitors of PI-3Kγ and PI-3Kδ are in development [142].

PPAR activators

Peroxisome proliferator-activated receptors (PPARs) are a family of ligand-activated nuclear hormone receptors belonging to the steroid receptor superfamily, and the three recognised subtypes PPAR-α, -γ and -δ are widely expressed. There is evidence that activation of PPAR-α and PPAR-δ may have anti-inflammatory and immunomodulatory effects. For example PPAR-γ agonists, such as troglitazone, inhibit the release of inflammatory cytokines from monocytes and induce apoptosis of T lymphocytes [143, 144], suggesting that they may have anti-inflammatory effects in COPD.

Strategies acting on structural cells

Mucoregulation

Mucus hypersecretion is commonly seen in cigarette smokers, but is not a stable feature of COPD. In individuals with COPD mucus hypersecretion is associated with more rapid decline in FEV_1 [19]. However, mucus hypersecretion following heavy cigarette smoking does not cause increased risk of airflow obstruction [20]. However, mucus hypersecretion may accelerate the decline in lung function in patients with COPD, however, by increasing the frequency of exacerbations. This suggests that reducing mucus hypersecretion may have therapeutic benefit, although suppression of the normal airway mucus secretion may be detrimental. There are several approaches to inhibiting mucus hypersecretion that are currently being explored [145] (Fig. 6). Mucus hypersecretion appears to be driven in COPD by the neutrophil inflammatory response, so that effective anti-inflammatory treatments would be expected to reduce mucus secretions [145, 146].

Epidermal growth factor (EGF) plays a critical role in airway mucus secretion from goblet cells and submucosal glands and appears to mediate the mucus secretory response to several secretagogues, including oxidative stress, cigarette smoke and inflammatory cytokies [17, 147]. Small molecule inhibitors of EGF receptor kinase, such as gefitinib, have now been developed for clinical use. There has been some concern about interstitial lung disease in some patients with small cell lung cancer treated with gefitinib, but it is not yet certain if this is related to EGF inhibition [148].

Another novel approach involves inhibition of calcium-activated chloride channels (CACC), which are important in mucus secretion from goblet cells. Activation of human hCLCA1 induces mucus secretion and mucus gene expression and may therefore be a target for inhibition. Small molecule inhibitors of CACC, such as niflumic acid and MSI 1956, have been developed [149].

Fibrosis

Transforming growth factor-β_1 (TGF-β_1) is highly expressed in airway epithelium and macrophages of small airways in patients with COPD [150, 151]. It is a potent inducer of fibrosis, partly *via* the release of the potent fibrogenic mediator connective tissue growth factor, and may be important in inducing the fibrosis and narrowing of peripheral airways (obstructive bronchiolitis) in COPD. MMP-9 may play a role in the activation of transforming growth factor-β_1 (TGF-β_1) [152, 153], as well as with release of chemotactic peptides and activation of α_1-antitrypsin; thus being involved closely with both proteolysis and fibrosis. TGF-β_1 also activates MMP-9, and this MMP-9 then further activates TGF-β_1, thus providing a link between small airway fibrosis and emphysema in COPD. MMP-9 may mediate proteolysis of TGF-β-binding protein (LTBP1), and this may be a mechanism for physiological release of TGF-β_1 [152]. TGF-β_1 also down-regulates β_2-adrenoceptors and thus may impair responses to β_2-agonists in peripheral airways [154]. Inhibition of TGF-β_1 signalling may therefore be a useful therapeutic strategy in COPD. Small molecule antagonists which inhibit TGF-β_1 receptor kinase are now in development [155], although the long-term safety of such drugs might be a problem, particularly as TGF-β_1 affects tissue repair and is a potent anti-inflammatory mediator.

Proteinase-activated receptor-2 (PAR-2) expression is widespread in the airways, and expression is similar in the central airways of smokers and non-smokers [156]. PAR-2 is a transmembrane receptor preferentially activated by trypsin and tryptase,

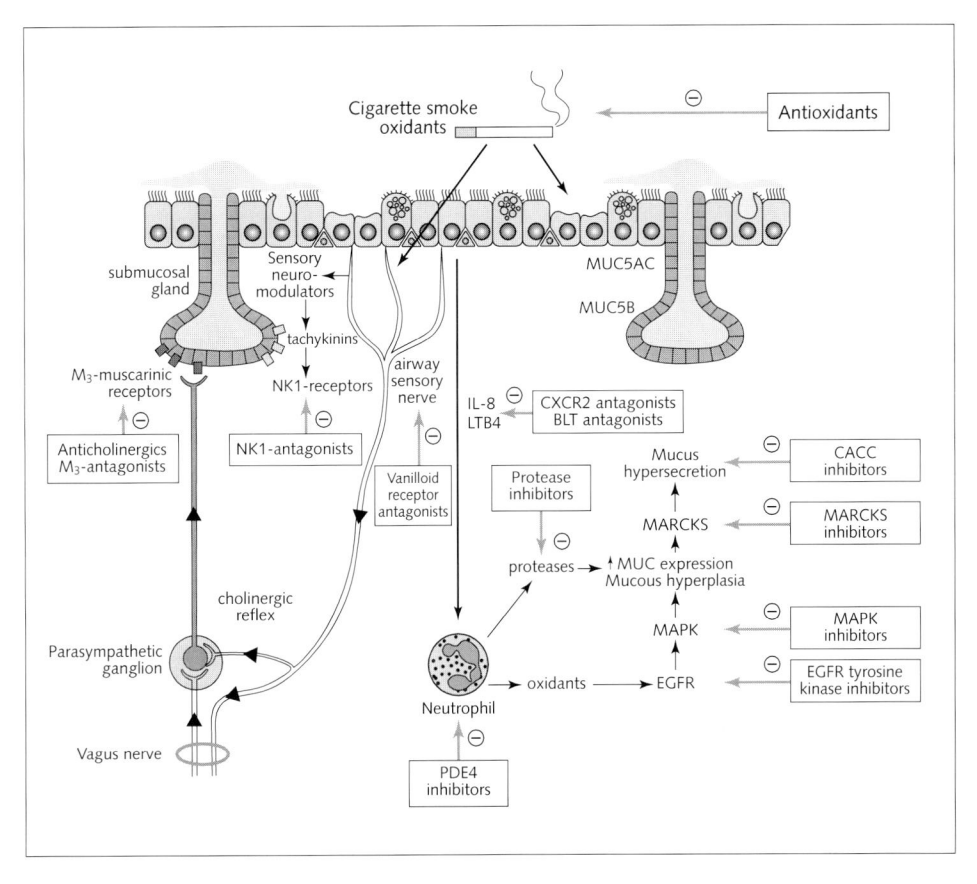

Figure 6
Inhibition of mucus hypersecretion. Abbreviations: CACC, calcium-activated chloride channel; EGFR, epidermal growth factor receptor; IL-8, interleukin-8; LTB_4, leukotriene B_4; MAPK, mitogen-activated protein kinase; MARCKS, myristoylated alanine-rich C kinase substrate; NK1, neurokinin 1; PDE4, phosphodiesterase 4.

and PARs play an important role in matrix remodelling, cell migration and proliferation and inflammation. PAR-2 may be involved in MMP-9 release from airway epithelial cells [157], proliferation of fibroblasts [158], and proliferation of airway smooth muscle [159]. Mast cell tryptase stimulates lung fibroblast proliferation *via* PAR-2 activation [159, 160]. However, a potential drawback for strategies to antagonise PAR-2 is that activation of epithelial PAR-2 causes bronchoprotection in the airways [161]. Vascular remodelling is recognised as an early feature of COPD, and fibroblast growth factors (FGF) 1 and 2 have been identified in vascular and airway smooth muscle, with elevated FGFR-1 in the intima of blood vessels [153]. Hence MMP-9, PAR-2 and TGF-β may be interrelated in causing fibrosis in COPD. In addition, there are links between inflammation, enzymes and fibrosis during airway remodelling; and these interactions could be useful targets for new drugs for COPD.

Proteases

Emphysema is due to an imbalance between proteases (that digest elastin and other structural proteins in the alveolar wall) and antiproteases, that protect against this attack [162]. Neutrophil elastase, a neutral serine protease, is a major constituent of lung elastolytic activity and also potently stimulates mucus secretion. In addition, neutrophil elastase induces IL-8 release from epithelial cells and therefore may perpetuate the inflammatory state. Although neutrophil elastase is likely to be the major mechanism mediating elastolysis in patients with α_1-AT deficiency, it may well not be the major elastolytic enzyme in smoking-related COPD, and it is important to consider other enzymes as targets for inhibition. Proteinase 3 is another neutral serine protease in neutrophils and may contribute to the elastolytic activity of these cells; while cathepsin G is another cysteine protease in neutrophils that has elastolytic activity. Cathepsins B, K, L and S are cysteine proteases that are released from macrophages and have elastolytic activity.

Matrix metalloproteinases (MMP) are a group of over 20 closely related endopeptidases that are capable of degrading all of the components of the extracellular matrix of lung parenchyma, including elastin, collagen, proteoglycans, laminin and fibronectin. They are produced by neutrophils, alveolar macrophages and airway epithelial cells [163]. Increased levels of collagenase (MMP-1) and gelatinase B (MMP-9) have been detected in bronchoalveolar lavage fluid of patients with emphysema [164]. BAL macrophages from patients with emphysema express more MMP-9 and MMP-1 than cells from control subjects, suggesting that these cells, rather than neutrophils, may be the major cellular source [165]. Alveolar macrophages also express a unique MMP, macrophage metalloelastase (MMP-12) [166]. MMP-12 knock-out mice do not develop emphysema and do not show the expected increases in lung macrophages after long-term exposure to cigarette smoke [167]. MMP-12 does not appear to play a major role in humans and MMP-9 is likely to be a major elastolytic enzyme in emphysema.

Counterbalancing these proteases are a range of antiproteases; α_1-AT, also known as α_1-protease inhibitor, being the major antiprotease in lung parenchyma. Inheritance of homozygous α_1-AT deficiency may result in severe emphysema, particularly in cigarette smokers, but this genetic disease accounts for < 1% of cases of COPD. α_1-AT is not the only antiprotease. Secretory leukocyte protease inhibitor (SLPI) may be the most important protective mechanism in the airways, being derived from airway epithelial cells and providing a local protective mechanism. Tissue inhibitors of metalloproteinases (TIMPs) counteract the effect of matrix metalloproteinases, while cystatins counteract the effect of cathepsins

This suggests that either inhibiting these proteolytic enzymes or increasing endogenous antiproteases may be beneficial and theoretically should prevent the progression of airflow obstruction in COPD [168, 169] (Fig. 7). One approach is to give endogenous antiproteases (α_1-AT, secretory leukoprotease inhibitor, elafin, tissue inhibitors of MMP), either in recombinant form or by viral vector gene delivery [170, 171]. These approaches are unlikely to be cost effective as large amounts of protein have to be delivered and gene therapy is unlikely to provide sufficient protein.

A more promising approach is to develop small molecule inhibitors of proteinases, particularly those that have elastolytic activity [172]. Small molecule inhibitors, such as ONO-5046 and FR901277, have been developed which have high potency [173, 174]. These drugs inhibit neutrophil elastase-induced lung injury in experimental animals, whether given by inhalation or systemically and also inhibit the other serine proteases released from neutrophils (cathepsin G and proteinase-3). Small molecule inhibitors of neutrophil elastase are now entering clinical trials, but there is concern that neutrophil elastase may not play a critical role in emphysema and that other proteases are more important in elastolysis. Inhibitors of elastolytic cysteine proteases, such as cathepsins K, S and L that are released from macrophages [175] are also in development [176]. Matrix metalloproteinases with elastolytic activity (such as MMP-9) may also be a target for drug development [177], although non-selective MMP inhibitors, such as marimastat, appear to have considerable side-effects. It is possible that side-effects could be reduced by increasing selectivity for specific MMPs or by targeting delivery to the lung parenchyma. MMP-9 is markedly over-expressed by alveolar macrophages from patients with COPD and is the major elastolytic enzyme released by these cells [178], so a selective inhibitor might be useful in the treatment of emphysema.

Lung regeneration

Since a major mechanism of airway obstruction in COPD is due to loss of elastic recoil due to proteolytic destruction of lung parenchyma, it seems unlikely that this could be reversible by drug therapy, although it might be possible to reduce the rate of progression by preventing the inflammatory and enzymatic disease process. Retinoic acid increases the number of alveoli in developing rats and, remarkably, reverses the histological and physiological changes induced by elastase treatment of adult rats [179, 180]. However, this is not observed in other species [181]. Retinoic acid activates retinoic acid receptors, which act as tran-

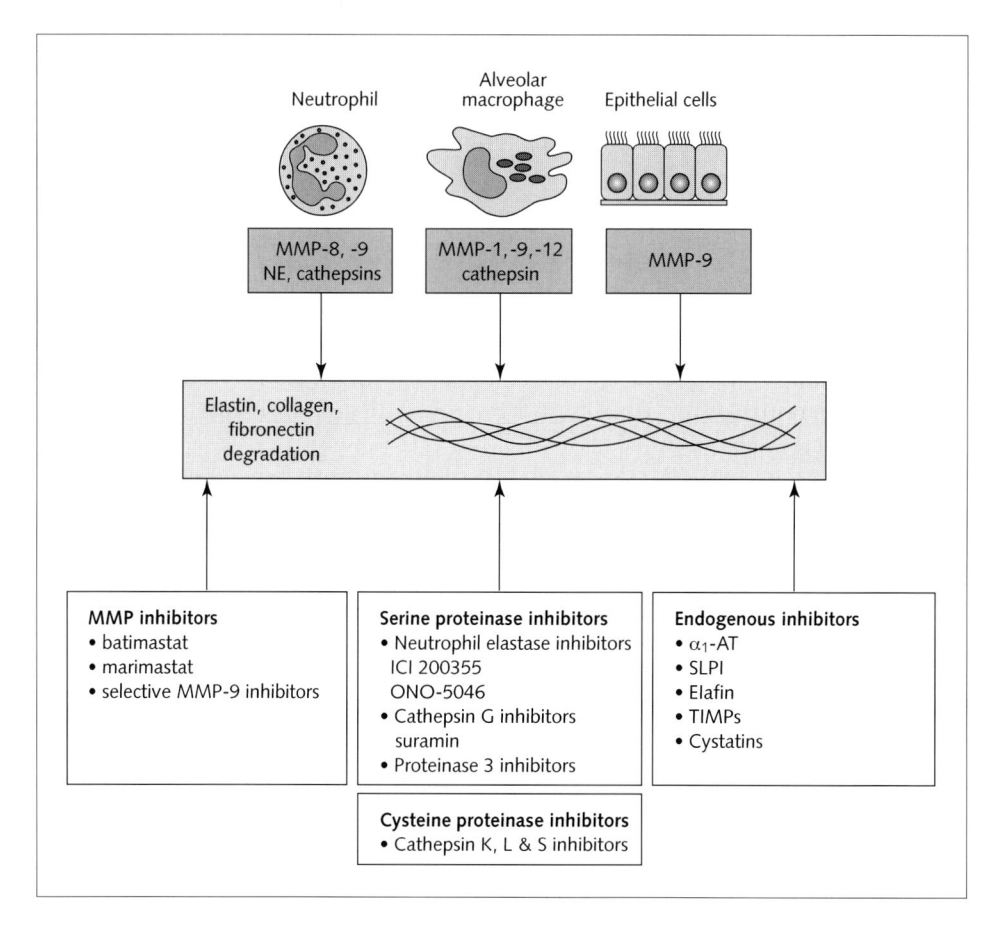

Figure 7
Protease inhibition. Matrix metalloproteases (MMP), neutrophil elastase (NE) and cathepsin catalyze the degradation of extracellular matrix proteins such as elastin, collagen and fibronectin. Endogenous protease inhibitors include α_1-antitrypsin (α_1-AT), secretory leukoprotease inhibitor (SLPI) and tissue inhibitor of matrix metalloproteinases (TIMP).

scription factors to regulate the expression of many genes involved in growth and differentiation. The molecular mechanisms involved and whether this can be extrapolated to humans is not yet known. Several retinoic acid receptor subtype agonists have now been developed that may have a greater selectivity for this effect and therefore a lower risk of side-effects. The receptor mediating the effect on alveoli appears to be the RAR-γ receptor. A short-term trial of all-trans-retinoic acid in patients with emphysema did not show any improvement in clinical parameters [182], but a longer study is currently underway. This approach is unlikely to be successful as adult human lung, unlike rat lung, has less potential for repair. Another approach to repairing damaged lung in emphysema is the use of stem cells to seed the lung [183]. Type 2 pneumocytes and Clara cells might be suitable for alveolar repair and this is currently an active area of research.

Future directions

New drugs for the treatment of COPD are greatly needed. While preventing and quitting smoking is the obvious preferred approach, this has proved to be very difficult in the majority of patients [184]. Furthermore, the contribution of other environmental factors (cooking fumes, pollutants, passive smoking, other inhaled toxins) and developmental changes in the lungs needs to be minimised [185]. It is important to identify the genetic factors that determine why only a minority of heavy smokers develop COPD [186], and identification of genes that predispose to the development of COPD may provide novel therapeutic targets. However, it will be difficult to demonstrate the efficacy of novel treatments on the rate of decline in lung function, since this requires large studies over three years. Hence, there is a need to develop novel outcome measures and surrogate biomarkers [187], such as analysis of sputum parameters (cells, mediators, enzymes) or exhaled condensates (lipid mediators, reactive oxygen species) [188]. It may also be important to more accurately define the presence of emphysema *versus* small airway obstruction using computerised tomography (CT) scans, as some drugs may be more useful for preventing emphysema, whereas others may be more effective against the small airway inflammatory and fibrotic processes. More research on the pathophysiology of COPD is urgently needed to aid the logical development of new therapies for this common and important disease, for which no effective preventative treatments currently exist.

References

1 Barnes PJ (2002) New treatments for COPD. *Nat Rev Drug Discov* 1: 437–446
2 Hansel TT, Barnes PJ (eds) (2001) *New Drugs for Asthma, Allergy and COPD*. Karger, Basel, Switzerland
3 Barnes PJ (2000) Chronic obstructive pulmonary disease. *N Engl J Med* 343: 269–280
4 Hogg JC (2001) Chronic obstructive pulmonary disease: an overview of pathology and pathogenesis. *Novartis Found Symp* 234: 4–19
5 Barnes PJ (2003) New concepts in chronic obstructive pulmonary disease. *Ann Rev Med* 54: 113–129
6 Barnes PJ (2000) Mechanisms in COPD: differences from asthma. *Chest* 117: 10S–14S

7   Saetta M, Turato G, Maestrelli P, Mapp CE, Fabbri LM (2001) Cellular and structural bases of chronic obstructive pulmonary disease. *Am J Respir Crit Care Med* 163: 1304–1309
8   Turato G, Zuin R, Saetta M (2001) Pathogenesis and pathology of COPD. *Respiration* 68: 4–19
9   Jeffery PK (2000) Comparison of the structural and inflammatory features of COPD and asthma. Giles F. Filley Lecture. *Chest* 117: 251S–260S
10 Calhoun WJ, Retamales I, Elliott WM, Meshi B, Coxson HO, Havashi S, Pate PD, Hogg JC (2002) More inflammation than lung in emphysema. *Am J Respir Crit Care Med* 165: 730B–7731
11 Jeffery PK (2001) Remodeling in asthma and chronic obstructive lung disease. *Am J Respir Crit Care Med* 164: S28–S38
12 Jeffery PK (2001) Lymphocytes, chronic bronchitis and chronic obstructive pulmonary disease. *Novartis Found Symp* 234: 149–161
13 Jeffery PK, Laitinen A, Venge P (2000) Biopsy markers of airway inflammation and remodelling. *Respir Med* 94 (Suppl F): S9–15
14 Corrin B (2000) *Pathology of the Lungs*. Churchill Livingstone, London
15 Reid L (1954) Pathology of chronic bronchitis. *Lancet* 1: 275–279
16 Reid L (1960) Measurement of the bronchial mucous gland layer: a diagnostic yardstick in chronic bronchitis. *Thorax* 6: 132–141
17 Takeyama K, Dabbagh K, Lee HM, Agusti C, Lausier JA, Ueki IF, Grattan KM, Nadel JA (1999) Epidermal growth factor system regulates mucin production in airways. *Proc Natl Acad Sci USA* 96: 3081–3086
18 Reid CJ, Gould S, Harris A (1997) Developmental expression of mucin genes in the human respiratory tract. *Am J Respir Cell Mol Biol* 17: 592–598
19 Vestbo J, Prescott E, Lange P (1996) Association of chronic mucus hypersecretion with FEV1 decline and chronic obstructive pulmonary disease morbidity. *Am J Respir Crit Care Med* 153: 1530–1535
20 Vestbo J and Lange P (2002) Can GOLD Stage 0 provide information of prognostic value in chronic obstructive pulmonary disease? *Am J Respir Crit Care Med* 166: 329–332
21 Ebert RV and Terracio MJ (1975) The bronchiolar epithelium in cigarette smokers. Observations with the scanning electron microscope. *Am Rev Respir Dis* 111: 4–11
22 Ebert RV and Hanks PB (1981) Mucus secretion by the epithelium of the bronchioles of cigarette smokers. *Br J Dis Chest* 75: 277–282
23 Cosio MG, Hale KA, Niewoehner DE (1980) Morphologic and morphometric effects of prolonged cigarette smoking on the small airways. *Am Rev Respir Dis* 122: 265–21
24 Macklem PT, Proctor DF, Hogg JC (1970) The stability of peripheral airways. *Respir Physiol* 8: 191–203
25 Jeffery PK (1998) Structural and inflammatory changes in COPD: a comparison with asthma. *Thorax* 53: 129–136

26 Saetta M, Turato G, Baraldo S, Zanin A, Braccioni F, Mapp CE, Maestrelli P, Cavallesco G, Papi A, Fabbri LM (2000) Goblet cell hyperplasia and epithelial inflammation in peripheral airways of smokers with both symptoms of chronic bronchitis and chronic airflow limitation. *Am J Respir Crit Care Med* 161: 1016–1021

27 Saetta M, Di Stefano A, Turato G, Facchini FM, Corbino L, Mapp CE, Maestrelli P, Ciaccia A, Fabbri LM (1998) CD8+ T-lymphocytes in peripheral airways of smokers with chronic obstructive pulmonary disease. *Am J Respir Crit Care Med* 157: 822–826

28 Lams BEA, Sousa AR, Rees PJ, Lee TH (1998) Immunopathology of the small-airway submucosa in smokers with and without chronic obstructive pulmonary disease. *Am J Respir Crit Care Med* 158: 1518–1523

29 Snider GL (1986) Chronic obstructive pulmonary disease – a continuing challenge. *Am Rev Respir Dis* 133: 942–944

30 Thurlbeck WM (1985) Chronic airflow obstruction: correlation of structure and function. In: TL Petty (ed): *Chronic Obstructive Pulmonary Disease*. 2nd ed. Marcel Dekker, New York, 129–203

31 Wright JL, Lawson LM, Pare PD, Wiggs BJ, Kennedy S, Hogg JC (1983) Morphology of peripheral airways in current smokers and ex-smokers. *Am Rev Respir Dis* 127: 474–477

32 Verbeken EK, Cauberghs M, Mertens I, Lauweryns JM, Van de Woestijne KP (1992) Tissue and airway impedance of excised normal, senile, and emphysematous lungs. *J Appl Physiol* 72: 2343–2353

33 Wright JL, Hobson JE, Wiggs B, Pare PD, Hogg JC (1988) Airway inflammation and peribronchiolar attachments in the lungs of nonsmokers, current and ex-smokers. *Lung* 166: 277–286

34 Niewoehner DE, Kleinerman J, Rice DB (1974) Pathologic changes in the peripheral airways of young cigarette smokers. *N Engl J Med* 291: 755–758

35 Reynolds HY (1987) Bronchoalveolar lavage. *Am Rev Respir Dis* 135: 250–263

36 Matsuba K, Thurlbeck WM (1972) The number and dimensions of small airways in emphysematous lungs. *Am J Pathol* 67: 265–275

37 Bignon J, Khoury F, Even P, Andre J, Brouet, G. (1969) Morphometric study in chronic obstructive bronchopulmonary disease. Pathologic, clinical, and physiologic correlations. *Am Rev Respir Dis* 99: 669–695

38 Hogg JC, Macklem PT, Thurlbeck WM (1968) Site and nature of airway obstruction in chronic obstructive lung disease. *N Engl J Med* 278: 1355–1360

39 Mitchell RS, Stanford RE, Johnson JM, Silvers GW, Dart G, George MS (1976) The morphologic features of the bronchi, bronchioles, and alveoli in chronic airway obstruction: a clinicopathologic study. *Am Rev Respir Dis* 114: 137–145

40 Hale KA, Ewing SL, Gosnell BA, Niewoehner DE (1984) Lung disease in long-term cigarette smokers with and without chronic air-flow obstruction. *Am Rev Respir Dis* 130: 716–721

41 Jamal K, Cooney TP, Fleetham JA, Thurlbeck WM (1984) Chronic bronchitis. Correla-

tion of morphologic findings to sputum production and flow rates. *Am Rev Respir Dis* 129: 719–722
42   Saetta M, Ghezzo H, Kim WD, King M, Angus GE, Wang NS, Cosio MG (1985) Loss of alveolar attachments in smokers. A morphometric correlate of lung function impairment. *Am Rev Respir Dis* 132: 894–900
43   Finkelstein R, Ma HD, Ghezzo H, Whittaker K, Fraser RS, Cosio MG (1995) Morphometry of small airways in smokers and its relationship to emphysema type and hyperresponsiveness. *Am J Respir Crit Care Med* 152: 267–276
44   Linhartova A, Anderson AEJ, Foraker AG (1977) Further observations on luminal deformity and stenosis of nonrespiratory bronchioles in pulmonary emphysema. *Thorax* 32: 50–53
45   Anderson AEJ and Foraker AG (1962) Relative dimensions of bronchioles and parenchymal spaces in lungs from normal subjects and emphysematous patients. *Am J Med* 32: 218–226
46   Gelb AF, Hogg JC, Muller NL, Schein MJ, Kuei J, Tashkin DP, Epstein JD, Kollin J, Green RH, Zamel N et al (1996) Contribution of emphysema and small airways in COPD. *Chest* 109: 353–359
47   Leopold JG, Goeff J (1957) Centrilobular form of hypertrophic emphysema and its relation to chronic bronchitis. *Thorax* 12: 219–235
48   Cosio MG, Cosio Piqueras MG (2000) Pathology of emphysema in chronic obstructive pulmonary disease. *Monaldi Arch Chest Dis* 55: 124–129
49   Gillooly M, Lamb D (1993) Microscopic emphysema in relation to age and smoking habit. *Thorax* 48: 491–495
50   Lamb D, McLean A, Gillooly M, Warren PM, Gould GA, MacNee W (1993) Relation between distal airspace size, bronchiolar attachments, and lung function. *Thorax* 48: 1012–1017
51   Finkelstein R, Fraser RS, Ghezzo H, Cosio MG (1995) Alveolar inflammation and its relation to emphysema in smokers. *Am J Respir Crit Care Med* 152: 1666–1672
52   Eidelman D, Saetta MP, Ghezzo H, Wang NS, Hoidal JR, King M, Cosio MG (1990) Cellularity of the alveolar walls in smokers and its relation to alveolar destruction. Functional implications. *Am Rev Respir Dis* 141: 1547–1552
53   Saetta M, Baraldo S, Corbino L, Turato G, Braccioni F, Rea F, Cavallesco G, Tropeano G, Mapp CE, Maestrelli P et al (1999) CD8+ve cells in the lungs of smokers with chronic obstructive pulmonary disease. *Am J Respir Crit Care Med* 160: 711–717
54   Lang MR, Fiaux GW, Gillooly M, Stewart JA, Hulmes DJ, Lamb D (1994) Collagen content of alveolar wall tissue in emphysematous and non-emphysematous lungs. *Thorax* 49: 319–326
55   Sekhon HS, Wright JL, Churg A (1994) Cigarette smoke causes rapid cell proliferation in small airways and associated pulmonary arteries. *Am J Physiol* 267: L557–L563
56   Peinado VI, Barbera JA, Abate P, Ramirez J, Roca J, Santos S, Rodriguez-Roisin R (1999) Inflammatory reaction in pulmonary muscular arteries of patients with mild chronic obstructive pulmonary disease. *Am J Respir Crit Care Med* 159: 1605–1611

57 Peinado VI, Barbera JA, Ramirez J, Gomez FP, Roca J, Jover L, Gimferrer JM, Rodriguez-Roisin R (1998) Endothelial dysfunction in pulmonary arteries of patients with mild COPD. *Am J Physiol* 274: L908–L913

58 Barbera JA, Peinado VI, Santos S (2000) Pulmonary hypertension in COPD: old and new concepts. *Monaldi Arch Chest Dis* 55: 445–449

59 Dinh-Xuan AT, Higenbottam TW, Clelland CA, Pepke-Zaba J, Cremona G, Butt AY, Large SR, Wells FC, Wallwork J (1991) Impairment of endothelium-dependent pulmonary-artery relaxation in chronic obstructive lung disease. *N Engl J Med* 324: 1539–1547

60 Wright JL, Lawson L, Pare PD, Hooper R O, Peretz DI, Nelems JM, Schulzer M, Hogg JC (1983) The structure and function of the pulmonary vasculature in mild chronic obstructive pulmonary disease. The effect of oxygen and exercise. *Am Rev Respir Dis* 128: 702–707

61 Giaid A, Yanagisawa M, Langleben D, Michel RP, Levy R, Shennib H, Kimura S, Masaki T, Duguid WP, Stewart DJ (1993) Expression of endothelin-1 in the lungs of patients with pulmonary hypertension. *N Engl J Med* 328: 1732–1739

62 Sofia M, Mormile M, Faraone S, Carratu P, Alifano M, Di Benedetto G, Carratu L (1994) Increased 24-hour endothelin-1 urinary excretion in patients with chronic obstructive pulmonary disease. *Respiration* 61: 263–268

63 Roland M, Bhowmik A., Sapsford R J, Seemungal TA, Jeffries DJ, Warner TD, Wedzicha JA (2001) Sputum and plasma endothelin-1 levels in exacerbations of chronic obstructive pulmonary disease. *Thorax* 56: 30–35

64 Barbera JA, Riverola A, Roca J, Ramirez J., Wagner PD, Ros D, Wiggs BR, Rodriguez-Roisin R (1994) Pulmonary vascular abnormalities and ventilation-perfusion relationships in mild chronic obstructive pulmonary disease. *Am J Respir Crit Care Med* 149: 423–429

65 Bernard S, LeBlanc P, Whittom F, Carrier G, Jobin J, Belleau R, Maltais F (1998) Peripheral muscle weakness in patients with chronic obstructive pulmonary disease. *Am J Respir Crit Care Med* 158: 629–634

66 Schols AM, Slangen J, Volovics L, Wouters, EF (1998) Weight loss is a reversible factor in the prognosis of chronic obstructive pulmonary disease. *Am J Respir Crit Care Med* 157: 1791–1797

67 de Godoy I, Donahoe M., Calhoun WJ, Mancino J, and Rogers RM (1996) Elevated TNF-alpha production by peripheral blood monocytes of weight- losing COPD patients. *Am J Respir Crit Care Med* 153: 633–637

68 Schols AM, Buurman WA, Staal van den Brekel AJ, Dentener MA, Wouters EF (1996) Evidence for a relation between metabolic derangements and increased levels of inflammatory mediators in a subgroup of patients with chronic obstructive pulmonary disease. *Thorax* 51: 819–824

69 Schols AM, Creutzberg EC, Buurman WA, Campfield LA, Saris WH, Wouters EF (1999)

Plasma leptin is related to pro-inflammatory status and dietary intake in patients with chronic obstructive pulmonary disease. *Am J Respir Crit Care Med* 160: 1220–1226

70 Eid AA, Ionescu AA, Nixon LS, Lewis-Jenkins V, Matthews SB, Griffiths TL, Shale DJ (2001) Inflammatory response and body composition in chronic obstructive pulmonary disease. *Am J Respir Crit Care Med* 164: 1414–1418

71 Schols AM, Soeters PB, Mostert R, Pluymers RJ, Wouters EF (1995) Physiologic effects of nutritional support and anabolic steroids in patients with chronic obstructive pulmonary disease. A placebo-controlled randomized trial. *Am J Respir Crit Care Med* 152: 1268–1274

72 Souness JE, Aldous D, Sargent C (2000) Immunosuppressive and anti-inflammatory effects of cyclic AMP phosphodiesterase (PDE) type 4 inhibitors. *Immunopharmacology* 47: 127–162

73 Giembycz MA (2002) Development status of second generation PDE4 inhibitors for asthma and COPD: the story so far. *Monaldi Arch Chest Dis* 57: 48–64

74 Huang Z, Ducharme Y, MacDonald D, Robichaud A (2001) The next generation of PDE4 inhibitors. *Curr Opin Chem Biol* 5: 432–438

75 Spond J, Chapman R, Fine J, Jones H, Kreutner W, Kung TT, Minnicozzi M (2001) Comparison of PDE 4 inhibitors, rolipram and SB 207499 (ariflo), in a rat model of pulmonary neutrophilia. *Pulm Pharmacol Ther* 14: 157–164

76 Hatzelmann A and Schudt C (2001) Anti-inflammatory and immunomodulatory potential of the novel PDE4 inhibitor roflumilast *in vitro*. *J Pharmacol Exp Ther* 297: 267–279

77 Compton CH, Gubb J, Nieman R, Edelson J, Amit O, Bakst A, Ayres JG, Creemers JP, Schultze-Werninghaus G, Brambilla C, Barnes NC (2001) Cilomilast, a selective phosphodiesterase-4 inhibitor for treatment of patients with chronic obstructive pulmonary disease: a randomised, dose-ranging study. *Lancet* 358: 265–270

78 Timmer W, Leclerc V, Birraux G, Neuhauser M, Hatzelmann A, Bethke T, Wurst W (2002) The new phosphodiesterase 4 inhibitor roflumilast is efficacious in exercise-induced asthma and leads to suppression of LPS-stimulated TNF-alpha *ex vivo*. *J Clin Pharmacol* 42: 297–303

79 Giembycz MA (2002) 4D or not 4D – the emetogenic basis of PDE4 inhibitors uncovered? *Trends Pharmacol Sci* 23: 548

80 Muller T, Engels P, Fozard JR (1996) Subtypes of the type 4 cAMP phosphodiesterases: structure, regulation and selective inhibition. *Trends Pharmacol Sci* 17: 294–299

81 Lamontagne S, Meadows E, Luk P, Normandin D, Muise E, Boulet L, Pon DJ, Robichaud A, Robertson GS, Metters KM, Nantel F (2001) Localization of phosphodiesterase-4 isoforms in the medulla and nodose ganglion of the squirrel monkey. *Brain Res* 920: 84–96

82 Robichaud A, Stamatiou PB, Jin SL, Lachance N, MacDonald D, Laliberte F, Liu S, Huang Z, Conti M, Chan CC (2002) Deletion of phosphodiesterase 4D in mice shortens alpha(2)-adrenoceptor-mediated anesthesia, a behavioral correlate of emesis. *J Clin Invest* 110: 1045–1052

83 Jin SL, Conti M (2002) Induction of the cyclic nucleotide phosphodiesterase PDE4B is essential for LPS-activated TNF-alpha responses. *Proc Natl Acad Sci USA* 99: 7628–7633

84 Reid P (2002) Roflumilast Altana Pharma. *Curr Opin Invest Drugs* 3: 1165–1170

85 Repine JE, Bast A, Lankhorst I (1997) Oxidative stress in chronic obstructive pulmonary disease. Oxidative Stress Study Group. *Am J Respir Crit Care Med* 156: 341–357

86 MacNee W (2000) Oxidants/antioxidants and COPD. *Chest* 117: 303S–317S

87 Ichinose M, Sugiura H, Yamagata S, Koarai A, Shirato K (2000) Increase in reactive nitrogen species production in chronic obstructive pulmonary disease airways. *Am J Respir Crit Care Med* 162: 701–706

88 MacNee W (2001) Oxidative stress and lung inflammation in airways disease. *Eur J Pharmacol* 429: 195–207

89 Montuschi P, Collins JV, Ciabattoni G, Lazzeri N, Corradi M, Kharitonov SA, Barnes PJ (2000) Exhaled 8-isoprostane as an *in vivo* biomarker of lung oxidative stress in patients with COPD and healthy smokers. *Am J Respir Crit Care Med* 162: 1175–1177

90 Paredi P, Kharitonov SA, Leak D, Ward S, Cramer D, Barnes PJ (2000) Exhaled ethane, a marker of lipid peroxidation, is elevated in chronic obstructive pulmonary disease. *Am J Respir Crit Care Med* 162: 369–373

91 Dekhuijzen PN, Aben KK, Dekker I, Aarts LP, Wielders PL, van Herwaarden CL, Bast A (1996) Increased exhalation of hydrogen peroxide in patients with stable and unstable chronic obstructive pulmonary disease. *Am J Respir Crit Care Med* 154: 813–816

92 Pratico D, Basili S, Vieri M, Cordova C, Violi F, Fitzgerald GA (1998) Chronic obstructive pulmonary disease is associated with an increase in urinary levels of isoprostane F2alpha-III, an index of oxidant stress. *Am J Respir Crit Care Med* 158: 1709–1714

93 Grandjean EM, Berthet P, Ruffmann R, Leuenberger P (2000) Efficacy of oral long-term N-acetylcysteine in chronic bronchopulmonary disease: a meta-analysis of published double-blind, placebo-controlled clinical trials. Clin Ther 22: 209–221

94 Poole PJ, Black PN (2001) Oral mucolytic drugs for exacerbations of chronic obstructive pulmonary disease: systematic review. *BMJ* 322: 1271–1274

95 Cuzzocrea S, Riley DP, Caputi AP, Salvemini D (2001) Antioxidant therapy: a new pharmacological approach in shock, inflammation, and ischemia/reperfusion injury. *Pharmacol Rev* 53: 135–159

96 Culpitt SV, Rogers DF, Barnes PJ, Donnelly LE (2003) Resveratrol has a greater inhibitory effect than corticosteroid in inhibiting alveolar macrophages from COPD patients. *Am J Respir Crit Care Med* 167: A91–A91

97 Hobbs AJ, Higgs A, Moncada S (1999) Inhibition of nitric oxide synthase as a potential therapeutic target. *Annu Rev Pharmacol Toxicol* 39: 191–220

98 Hansel TT, Kharitonov SA, Donnelly LE, Erin EM, Currie MG, Moore WM, Manning PT, Recker DP, Barnes PJ (2003) A selective inhibitor of inducible nitric oxide synthase inhibits exhaled breath nitric oxide in healthy volunteers and asthmatics. *FASEB J* 17: 1298–1300

99 Hill AT, Bayley DL, Stockley RA (1999) The interrelationship of sputum inflammatory markers in patients with chronic bronchitis. *Am J Respir Crit Care Med* 160: 893–898

100 Hubbard RC, Fells G, Gadek J, Pacholok S, Humes J, Crystal RG (1991) Neutrophil accumulation in the lung in alpha 1-antitrypsin deficiency. Spontaneous release of leukotriene B4 by alveolar macrophages. *J Clin Invest* 88: 891–897

101 Yokomizo T, Kato K, Terawaki K, Izumi T, Shimizu T (2000) A second leukotriene B(4) receptor, BLT2. A new therapeutic target in inflammation and immunological disorders. *J Exp Med* 192: 421–432

102 Silbaugh SA, Stengel PW, Cockerham SL, Froelich LL, Bendele AM, Spaethe SM, Sofia M J, Sawyer JS, Jackson WT (2000) Pharmacologic actions of the second generation leukotriene B4 receptor antagonist LY29311: *in vivo* pulmonary studies. *Naunyn Schmiedebergs Arch Pharmacol* 361: 397–404

103 Crooks, SW, Bayley DL, Hill SL, Stockley RA (2000) Bronchial inflammation in acute bacterial exacerbations of chronic bronchitis: the role of leukotriene B4. *Eur Respir J* 15: 274–280

104 Beeh KM, Kornmann O, Buhl R, Culpitt SV, Giembycz MA, Barnes PJ (2003) Neutrophil chemotactic activity of sputum from patients with COPD: role of interleukin 8 and leukotriene B4. *Chest* 123: 1240–1247

105 Gompertz S, Stockley RA (2002) A randomized, placebo-controlled trial of a leukotriene synthesis inhibitor in patients with COPD. *Chest* 122: 289–294

106 Davenpeck KL, Berens KL, Dixon RA, Dupre B, Bochner BS (2000) Inhibition of adhesion of human neutrophils and eosinophils to P- selectin by the sialyl Lewis antagonist TBC1269: preferential activity against neutrophil adhesion *in vitro*. *J Allergy Clin Immunol* 105: 769–775

107 Noguera A, Batle S, Miralles C, Iglesias J, Busquets X, MacNee W, Agusti AG (2001) Enhanced neutrophil response in chronic obstructive pulmonary disease. *Thorax* 56: 432–437

108 Hay DWP, Sarau HM (2001) Interleukin-8 receptor antagonists in pulmonary diseases. *Curr Opin Pharm* 1: 242–247

109 Keatings VM, Collins PD, Scott DM, Barnes PJ (1996) Differences in interleukin-8 and tumour necrosis factor-alpha in induced sputum from patients with chronic obstructive pulmonary disease or asthma. *Am J Respir Crit Care Med* 153: 530–534

110 Yamamoto C, Yoneda T, Yoshikawa M, Fu A, Tokuyama T, Tsukaguchi K, Narita N (1997) Airway inflammation in COPD assessed by sputum levels of interleukin-8. *Chest* 112: 505–510

111 Pesci A, Balbi B, Majori M, Cacciani G, Bertacco S, Alciato P, Donner CF (1998) Inflammatory cells and mediators in bronchial lavage of patients with chronic obstructive pulmonary disease. *Eur Respir J* 12: 380–386

112 Yang XD, Corvalan JR, Wang P, Roy CM, Davis CG (1999) Fully human anti-interleukin-8 monoclonal antibodies: potential therapeutics for the treatment of inflammatory disease states. *J Leukocyte Biol* 66: 401–410

113 Richman-Eisenstat JB, Jorens PG, Hebert CA, Ueki I, Nadel JA (1993) Interleukin-8: an

important chemoattractant in sputum of patients with chronic inflammatory airway diseases. *Am J Physiol* 264: L413–L418

114 Rossi D, Zlotnik A (2000) The biology of chemokines and their receptors. *Annu Rev Immunol* 18: 217–242

115 Traves SL, Culpit SV, Russell RE, Barnes PJ, Donnelly LE (2002) Increased levels of the chemokines GROalpha and MCP-1 in sputum samples from patients with COPD. *Thorax* 57: 590–595

116 White JR, Lee JM, Young PR, Hertzberg RP, Jurewicz AJ, Chaikin MA, Widdowson K, Foley JJ, Martin LD, Griswold DE, Sarau HM (1998) Identification of a potent, selective non-peptide CXCR2 antagonist that inhibits interleukin-8-induced neutrophil migration. *J Biol Chem* 273: 10095–10098

117 Hay DW, Sarau HM (2001) Interleukin-8 receptor antagonists in pulmonary diseases. *Curr Opin Pharmacol* 1: 242–247

118 Capelli A, Stefano A, Gnemmi I, Balbo P, Cerutti CG, Balbi B, Lusuardi M, Donner CF (1999) Increased MCP-1 and MIP-1beta in bronchoalveolar lavage fluid of chronic bronchitics. *Eur Respir J* 14: 160–165

119 de Boer WI, Sont JK, van Schadewijk A, Stolk J, van Krieken JH, Hiemstra PS (2000) Monocyte chemoattractant protein 1, interleukin 8, and chronic airways inflammation in COPD. *J Pathol* 190: 619–626

120 Saetta M, Mariani M, Panina-Bordignon P, Turato G, Buonsanti C, Baraldo S, Bellettato CM, Papi A, Corbetta L, Zuin R et al (2002) Increased expression of the chemokine receptor CXCR3 and its ligand CXCL10 in peripheral airways of smokers with chronic obstructive pulmonary disease. *Am J Respir Crit Care Med* 165: 1404–1409

121 Aaron SD, Angel JB, Lunau M, Wright K, Fex C, Le Saux N, Dales RE (2001) Granulocyte inflammatory markers and airway infection during acute exacerbation of chronic obstructive pulmonary disease. *Am J Respir Crit Care Med* 163: 349–355

122 Langen RC, Schols AM, Kelders MC, Wouters EF, Janssen-Heininger YM (2001) Inflammatory cytokines inhibit myogenic differentiation through activation of nuclear factor-kappaB. *FASEB J* 15: 1169–1180

123 Vernooy JH, Kucukaycan, M, Jacobs JA, Chavannes NH, Buurman WA, Dentener MA, Wouters EF (2002) Local and systemic inflammation in patients with chronic obstructive pulmonary disease: soluble tumor necrosis factor receptors are increased in sputum. *Am J Respir Crit Care Med* 166: 1218–1224

124 Markham A, Lamb HM (2000) Infliximab: a review of its use in the management of rheumatoid arthritis. *Drugs* 59: 1341–1359

125 Jarvis B, Faulds D (1999) Etanercept: a review of its use in rheumatoid arthritis. *Drugs* 57: 945–966

126 Barlaam B, Bird TG, Lambert-Van Der Brempt, C, Campbell D, Foster SJ, Maciewicz R (1999) New alpha-substituted succinate-based hydroxamic acids as TNFalpha convertase inhibitors. *J Med Chem* 42: 4890–4908

127 Rabinowitz MH, Andrews RC, Becherer JD, Bickett DM, Bubacz DG, Conway JG, Cowan D. J, Gaul M, Glennon K, Lambert MH et al (2001) Design of selective and sol-

uble inhibitors of tumor necrosis factor-alpha converting enzyme (TACE). *J Med Chem* 44: 4252–4267

128 Lee JC, Kumar S, Griswold DE, Underwood DC, Votta BJ, Adams JL (2000) Inhibition of p38 MAP kinase as a therapeutic strategy. *Immunopharmacology* 47: 185–201

129 Takanashi S, Hasegawa Y, Kanehira Y, Yamamoto K, Fujimoto K, Satoh K, Okamura K (1999) Interleukin-10 level in sputum is reduced in bronchial asthma, COPD and in smokers. *Eur Respir J* 14: 309–314

130 Fedorak RN, Gangl A, Elson CO, Rutgeerts P, Schreiber S, Wild G, Hanauer SB, Kilian A, Cohard M, LeBeaut A, Feagan B (2000) Recombinant human interleukin 10 in the treatment of patients with mild to moderately active Crohn's disease. The Interleukin 10 Inflammatory Bowel Disease Cooperative Study Group. *Gastroenterology* 119: 1473–1482

131 Di Stefano A, Caramori G, Capelli A, Lusuardi M, Gnemmi I, Ioli F, Chung KF, Donner C F., Barnes PJ, Adcock IM (2002) Increased expression of NF-kB in bronchial biopsies from smokers and patients with COPD. *Eur Respir J* 20: 556–563

132 Caramori G, Romagnoli M, Casolari P, Bellettato C, Casoni G, Boschetto P, Fan Chung K, Barnes PJ, Adcock IM, Ciaccia A et al (2003) Nuclear localisation of p65 in sputum macrophages but not in sputum neutrophils during COPD exacerbations. *Thorax* 58: 348–351

133 Delhase M, Li N, Karin M (2000) Kinase regulation in inflammatory response. *Nature* 406: 367–368

134 Castro AC, Dang LC, Soucy F, Grenier L, Mazdiyasni H, Hottelet M, Parent L, Pien C, Palombella V, Adams J (2003) Novel IKK inhibitors: beta-carbolines. *Bioorg Med Chem Lett* 13: 2419–2422

135 Culpitt SV, Rogers DF, Shah P, De Matos C, Russell RE, Donnelly LE, Barnes PJ (2003) Impaired inhibition by dexamethasone of cytokine release by alveolar macrophages from patients with chronic obstructive pulmonary disease. *Am J Respir Crit Care Med* 167: 24–31

136 Nasuhara Y, Adcock IM, Catley M, Barnes PJ, Newton R (1999) Differential IkappaB kinase activation and IkappaBalpha degradation by interleukin-1beta and tumor necrosis factor-alpha in human U937 monocytic cells. Evidence for additional regulatory steps in kappaB-dependent transcription. *J Biol Chem* 274: 19965–19972

137 Johnson GL, Lapadat R (2002) Mitogen-activated protein kinase pathways mediated by ERK, JNK, and p38 protein kinases. *Science* 298: 1911–1912

138 Meja KK, Seldon PM, Nasuhara Y, Ito K, Barnes PJ, Lindsay MA, Giembycz MA (2000) p38 MAP kinase and MKK-1 co-operate in the generation of GM-CSF from LPS-stimulated human monocytes by an NF-kappa B-independent mechanism. *Br J Pharmacol* 131: 1143–1153

139 Carter AB, Monick MM, Hunninghake GW (1999) Both Erk and p38 kinases are necessary for cytokine gene transcription. *Am J Respir Cell Mol Biol* 20: 751–758

140 Underwood DC, Osborn RR, Bochnowicz S, Webb EF, Rieman DJ, Lee JC, Romanic AM, Adams JL, Hay DW, Griswold DE (2000) SB 239063, a p38 MAPK inhibitor,

reduces neutrophilia, inflammatory cytokines, MMP-9, and fibrosis in lung. *Am J Physiol Lung Cell Mol Physiol* 279: L895–L902

141 Sasaki T, Irie-Sasaki J, Jones RG, Oliveira-dos-Santos AJ, Stanford WL, Bolon B, Wakeham A, Itie A, Bouchard D, Kozieradzki I et al (2000) Function of PI3Kgamma in thymocyte development, T cell activation, and neutrophil migration. *Science* 287: 1040–1046

142 Ward S, Sotsios Y, Dowden J, Bruce I, Finan P (2003) Therapeutic potential of phosphoinositide 3-kinase inhibitors. *Chem Biol* 10: 207–213

143 Jiang C, Ting AT, Seed B (1998) PPAR-gamma agonists inhibit production of monocyte inflammatory cytokines. *Nature* 391: 82–86

144 Harris SG, Phipps RP (2002) Induction of apoptosis in mouse T cells upon peroxisome proliferator-activated receptor gamma (PPAR-gamma) binding. *Adv Exp Med Biol* 507: 421–425

145 Barnes PJ (2002) Current and future therapies for airway mucus hypersecretion. *Novartis Found Symp* 248: 237–249

146 Goswami SK, Kivity S, Marom Z (1990) Erythromycin inhibits respiratory glycoconjugate secretion from human airways *in vitro*. *Am Rev Respir Dis* 141: 72–78

147 Nadel JA, Burgel PR (2001) The role of epidermal growth factor in mucus production. *Curr Opin Pharmacol* 1: 254–258

148 Inoue A, Saijo Y, Maemondo M, Gomi K, Tokue Y, Kimura Y, Ebina M, Kikuchi T, Moriya T, Nukiwa T (2003) Severe acute interstitial pneumonia and gefitinib. *Lancet* 361: 137–139

149 Zhou Y, Shapiro M, Dong Q, Louahed J, Weiss C, Wan S, Chen Q, Dragwa C, Savio D, Huang M et al (2002) A calcium-activated chloride channel blocker inhibits goblet cell metaplasia and mucus overproduction. *Novartis Found Symp* 248: 150–165

150 de Boer WI, van Schadewijk A, Sont JK, Sharma HS, Stolk J, Hiemstra PS, van Krieken JH (1998) Transforming growth factor beta1 and recruitment of macrophages and mast cells in airways in chronic obstructive pulmonary disease. *Am J Respir Crit Care Med* 158: 1951–1957

151 Takizawa H, Tanaka M, Takami K, Ohtoshi T, Ito K, Satoh M, Okada Y, Yamasawa F, Nakahara K, Umeda A (2001) Increased expression of transforming growth factor-beta1 in small airway epithelium from tobacco smokers and patients with chronic obstructive pulmonary disease (COPD). *Am J Respir Crit Care Med* 163: 1476–1483

152 Dallas S L, Rosser JL, Mundy GR, Bonewald LF (2002) Proteolysis of latent transforming growth factor-beta (TGF-beta)-binding protein-1 by osteoclasts. A cellular mechanism for release of TGF-beta from bone matrix. *J Biol Chem* 277: 21352–21360

153 Kranenburg AR, de Boer WI, van Krieken JH, Mooi WJ, Walters JE, Saxena PR, Sterk PJ, Sharma HS (2002) Enhanced expression of fibroblast growth factors and receptor FGFR-1 during vascular remodeling in chronic obstructive pulmonary disease. *Am J Respir Cell Mol Biol* 27: 517–525

154 Mak JC, Rousell J, Haddad EB, Barnes PJ (2000) Transforming growth factor-beta1

inhibits beta2-adrenoceptor gene transcription. *Naunyn Schmiedebergs Arch Pharmacol* 362: 520–525

155 Yakymovych I, Engstrom U, Grimsby S, Heldin CH, Souchelnytskyi S (2002) Inhibition of transforming growth factor-beta signaling by low molecular weight compounds interfering with ATP- or substrate-binding sites of the TGF beta type I receptor kinase. *Biochemistry* 41: 11000–11007

156 Miotto D, Hollenberg MD, Bunnett NW, Papi A, Braccioni F, Boschetto P, Rea F, Zuin A, Geppetti P, Saetta M et al (2002) Expression of protease activated receptor-2 (PAR-2) in central airways of smokers and non-smokers. *Thorax* 57: 146–151

157 Vliagoftis H, Schwingshackl A, Milne CD, Duszyk M, Hollenberg MD, Wallace JL, Befus A D, Moqbe R (2000) Proteinase-activated receptor-2-mediated matrix metalloproteinase-9 release from airway epithelial cells. *J Allergy Clin Immunol* 106: 537–545

158 Frungieri MB, Weidinger S, Meineke V, Kohn FM, Mayerhofer A (2002) Proliferative action of mast-cell tryptase is mediated by PAR2, COX2, prostaglandins, and PPARgamma: Possible relevance to human fibrotic disorders. *Proc Natl Acad Sci USA* 99: 15072–15077

159 Berger P, Perng DW, Thabrew H, Compton SJ, Cairns JA, McEuen AR., Marthan R, Tunon De Lara JM, Walls AF (2001) Tryptase and agonists of PAR-2 induce the proliferation of human airway smooth muscle cells. *J Appl Physiol* 91: 1372–1379

160 Akers IA Parsons M, Hill M R, Hollenberg MD, Sanjar S, Laurent GJ, McAnulty RJ (2000) Mast cell tryptase stimulates human lung fibroblast proliferation *via* protease-activated receptor-2. *Am J Physiol Lung Cell Mol Physiol* 278: L193–L201

161 Cocks TM, Fong B, Chow JM, Anderson GP, Frauman AG, Goldie RG, Henry PJ, Carr MJ, Hamilton JR, Moffatt JD (1999) A protective role for protease-activated receptors in the airways. *Nature* 398: 156–160

162 Stockley RA (1994) The role of proteinases in the pathogenesis of chronic bronchitis. *Am J Respir Crit Care Med* 150: S109–S113

163 Shapiro SD (1994) Elastolytic metalloproteinases produced by human mononuclear phagocytes. Potential roles in destructive lung disease. *Am J Respir Crit Care Med* 150: S160–S164

164 Finlay GA, Russell KJ, McMahon KJ, D'arcy EM, Masterson JB, FitzGerald MX, O'Connor, CM (1997) Elevated levels of matrix metalloproteinases in bronchoalveolar lavage fluid of emphysematous patients. *Thorax* 52: 502–506

165 Finlay GA, O'Driscoll LR, Russell KJ, D'arcy EM, Masterson JB, FitzGerald MX, O'Connor CM (1997) Matrix metalloproteinase expression and production by alveolar macrophages in emphysema. *Am J Respir Crit Care Med* 156: 240–247

166 Shapiro SD, Kobayashi DK, Ley TJ (1993) Cloning and characterization of a unique elastolytic metalloproteinase produced by human alveolar macrophages. *J Biol Chem* 268: 23824–23829

167 Hautamaki RD, Kobayashi DK, Senior RM, Shapiro SD (1997) Requirement for macrophage metalloelastase for cigarette smoke-induced emphysema in mice. *Science* 277: 2002–2004

168 Stockley RA (1999) Neutrophils and protease/antiprotease imbalance. *Am J Respir Crit Care Med* 160: S49–S52
169 Shapiro SD, Senior RM (1999) Matrix metalloproteinases. Matrix degradation and more. *Am J Respir Cell Mol Biol* 20: 1100–1102
170 Seersholm N, Wencker M, Banik N, Viskum K, Dirksen A, Kok-Jensen A, Konietzko N (1997) Does alpha1-antitrypsin augmentation therapy slow the annual decline in FEV1 in patients with severe hereditary alpha1-antitrypsin deficiency? Wissenschaftliche Arbeitsgemeinschaft zur Therapie von Lungenerkrankungen (WATL) alpha1-AT study group. *Eur Respir J* 10: 2260–2263
171 McElvaney NG, Doujaiji B, Moan MJ, Burnham MR, Wu MC, Crystal RG (1993) Pharmacokinetics of recombinant secretory leukoprotease inhibitor aerosolized to normals and individuals with cystic fibrosis. *Am Rev Respir Dis* 148: 1056–1060
172 Luisetti M, Sturani C, Sella D, Madonini E, Galavotti V, Bruno G, Peona V, Kucich U, Dagnino G, Rosenbloom J et al (1996) MR889, a neutrophil elastase inhibitor, in patients with chronic obstructive pulmonary disease: a double-blind, randomized, placebo- controlled clinical trial. *Eur Respir J* 9: 1482–1486
173 Kawabata K, Suzuki M, Sugitani M, Imaki K, Toda M, Miyamoto T (1991) ONO-5046, a novel inhibitor of human neutrophil elastase. *Biochem Biophys Res Commun* 177: 814–820
174 Fujie K, Shinguh Y, Yamazaki A, Hatanaka H, Okamoto M, Okuhara M (1999) Inhibition of elastase-induced acute inflammation and pulmonary emphysema in hamsters by a novel neutrophil elastase inhibitor FR901277. *Inflamm Res* 48: 160–167
175 Punturieri A, Filippov S, Allen E, Caras I, Murray R, Reddy V, Weiss SJ (2000) Regulation of elastinolytic cysteine proteinase activity in normal and cathepsin K-deficient human macrophages. *J Exp Med* 192: 789–799
176 Leung-Toung R, Li W, Tam TF, Karimian K (2002) Thiol-dependent enzymes and their inhibitors: a review. *Curr Med Chem* 9: 979–1002
177 Cawston TE (1996) Metalloproteinase inhibitors and the prevention of connective tissue breakdown. *Pharmacol Ther* 70: 163–182
178 Russell RE, Culpitt SV, DeMatos C, Donnelly L, Smith M, Wiggins J, Barnes PJ (2002) Release and activity of matrix metalloproteinase-9 and tissue inhibitor of metalloproteinase-1 by alveolar macrophages from patients with chronic obstructive pulmonary disease. *Am J Respir Cell Mol Biol* 26: 602–609
179 Massaro GD, Massaro D (1997) Retinoic acid treatment abrogates elastase-induced pulmonary emphysema in rats. *Nat Med* 3: 675–677
180 Belloni PN, Garvin L, Mao CP, Bailey-Healy I, Leaffer D (2000) Effects of all-transretinoic acid in promoting alveolar repair. *Chest* 117: 235S–241S
181 Lucey EC, Goldstein RH, Breuer R, Rexer BN, Ong DE, Snider GL (2003) Retinoic acid does not affect alveolar septation in adult FVB mice with elastase-induced emphysema. *Respiration* 70: 200–205
182 Mao JT, Goldin JG, Dermand J, Ibrahim G, Brown MS, Emerick A, McNitt-Gray MF,

Gjertson DW, Estrada F, Tashkin DP, Roth MD (2002) A pilot study of all-trans-retinoic acid for the treatment of human emphysema. *Am J Respir Crit Care Med* 165: 718–723
183 Otto WR (2002) Lung epithelial stem cells. *J Pathol* 197: 527–535
184 Lancaster T, Stead L, Silagy C, Sowden A (2000) Effectiveness of interventions to help people stop smoking: findings from the Cochrane Library. *Br Med J* 321: 355–358
185 Smith KR (2000) National burden of disease in India from indoor air pollution. *Proc Natl Acad Sci USA* 97: 13286–13293
186 Sandford AJ, Silverman EK (2002) Chronic obstructive pulmonary disease. 1: Susceptibility factors for COPD the genotype-environment interaction. *Thorax* 57: 736–741
187 Jones PW, Mahler DA (2002) Key outcomes in COPD: health-related quality of life. Proceedings of an expert round table held July 20–22, 2001 in Boston, Massachusetts, USA. *Eur Respir Rev* 12: 57–107
188 Kharitonov SA, Barnes PJ (2001) Exhaled markers of pulmonary disease. *Am J Respir Crit Care Med* 163: 1693–1722

Index

ahhesion molecule 156, 204
adhesion molecule blocker 204
advanced physiologic techniques 39
airflow limitation 21–23
airway inflammation 22–24
ambroxol 112
amplification of inflammation, oxidative stress 66
anticholinergics 109
anti-inflammatory treatment 137
antioxidants 62, 68, 107, 109, 199
antioxidants as therapy for COPD 68
α_1-antitrypsin (α_1-AT) 76, 79, 86, 210
α_1-AT, animal model 86
α_1-AT, deficiency 79
α_1-AT, elastase 79
apoptosis 67, 112
association studies, genetics 4
asthma 123, 137
azurophil granules, Chediak-Higashi syndrome 82
azurophil granules, exocytosis 79
azurophil granules, neutrophil elastase 79

beige mouse 83
BLT_1 antagonist 107
bromhexine 112
bulla 50, 55

C_4 141
calcium-activated chloride channel (CLCA) 108, 111, 208

calcium-activated potassium channel 110
candidate gene 6, 7
cannabinoids 110
capsazepine 110
carbocysteine 112
carbon monoxide gas transfer 38
cardiovascular effect 160
cathepsin B 85
cathepsin B, animal model 85
cathepsin B, bronchial disease 85
cathepsin B, emphysema 85
cathepsin C 84
cathepsin C, knockout mouse 84
cathepsin G 83, 85, 107
cathepsin G, animal model 85
cathepsin G, Chediak-Higashi syndrome 83
cathepsin G, mucous gland hyperplasia 85
cathepsin G, mucus 85
cathepsin L 86
cathepsin S 86
celecoxib 106
cell surface proteinase 81
C-fibre 109
Chediak-Higashi syndrome 82, 83
chemokine inhibitor 204
chest radiograph (CXR) 47, 48
cholinergic nerve 109, 110
chronic bronchitis 101, 106, 190
chronic inflammation 190
cigarette smoke 22, 24, 86, 140
cigarette smoke, animal model 86
cigarette smoke, MMP-12 86

cigarette smoke, proteinase inhibitor 86
ciliary function 76, 78
collagenase 88, 89, 131
computed tomography (CT) 47–60
computed tomography, airways disease 53, 54
computed tomography, assessment for surgery 56
computed tomography, attenuation values 48
computed tomography, conventional CT 49
computed tomography, density mask 52
computed tomography, detection of early disease 56
computed tomography, emphysema 49–53
computed tomography, ground glass attenuation 54, 55
computed tomography, high resolution CT (HRCT) 49
computed tomography, mean lung density 52
computed tomography, monitoring progress 56
computed tomography, parenchymal micronodules 54, 55
computed tomography, quantitative analysis of emphysema 52, 53
computed tomography, respiratory bronchiolitis-interstitial lung disease (RB-ILD) 54
computed tomography, spiral CT 49
condensate 137
COPD phenotype 1
corticosteroid resistance 67
COX-2 inhibitors 106
CXCR2 antagonist 107
cystatin C 86
cytokines 124

D_4 141
depression 161
dipeptidyl peptidase 84
dithiothreitol 124
dyspnoe 169

E_4 141
ebselen 68

education 177
eicosanoid 140
elastase 76, 78, 79, 81–83, 86, 107, 108
emphysema 21, 23–27, 49, 53, 75–78, 85, 126, 196
emphysema, animal models 76
emphysema, CT features 49–53
emphysema, elastase 78
emphysema, induction of 76
emphysema, proteinase 76
emphysema, proteinase-antiproteinase theory 75
emphysema, subclinical 78
endocrine disruption 160
eosinophil cationic protein (ECP) 123
eosinophil peroxidase (EPO) 123
epidermal growth factor (EGF-R) 108, 111, 208
epithelial neutrophil activating protein 78 (ENA-78) 130
epoxygenase 106
erythromycin 113
eterocoxib 106
exacerbation 78, 126
exacerbation, bacteria 78
exacerbation, elastase 78
excercise tolerance 169
excercise training 172
exhaled breath condensate (EBC) 140
exhaled breath temperature, bronchial blood flow 141
exhaled carbon monoxide (CO) 65, 147
exhaled ethane 65, 140
exhaled gas 137
exhaled markers of oxidative stress 65, 66
exhaled NO 137
exocytosis, Chediak-Higashi syndrome 83
exocytosis of azurophil granules 83

F_4 141
fibrosis 208
forced expiratory volume in 1 second (FEV_1) 31

gelatinase B 88, 89, 131
genetic background 160
genomic control 5
genomics 150
glucocorticosteroids 106
goblet cell, apoptosis 112
goblet cell, hyperplasia 103, 104, 106, 107, 111–113
growth related oncogene-α (GRO-α) 124

Haemophilus influenzae, proteinase 90
hCLCA1 206
histone deacetylase-2 (HDAC2) 67
house dust mite, proteinase 90
human neutrophil lipocalin (HNL) 122
hydrocarbon 140
hydrogen peroxide (H_2O_2) 61, 140
hydroxyl radical ($^{\cdot}OH$) 61

IL-8 78, 79, 124
IL-8, lung secretions 79
IL-8, neutrophil recruitment 79
IL-10 125, 205
increased oxidative stress 65, 66
inert gas technique 34
inspiratory capacity, dynamic hyperinflation 39
8-isoprostane 140

large conductance calcium-activated potassium channel 110
leukotriene 78, 79, 130, 140, 141, 202
leukotriene (LT) B4 78, 79, 130, 141
leukotriene inhibitor 202
linkage analysis 3
linkage disequilibrium 5
LTB_4 130, 141
LTB4, bronchoalveolar lavage fluid 79
LTB4, elastase 79
LTB4, macrophage 79
LTB4, neutrophil recruitment 79
lung regeneration 211
lung transplantation 55

lung volume reduction surgery (LVRS) 55
lung volumes, plethysmographic 37

macrolide antibiotics 113
macrophage 86, 128
macrophage, proteinases 86
macrophage metalloelastase (MMP-12) 86
macrophage metalloelastase (MMP-12), knock-out mouse 86
mast cell 85
mast cell derived chemotryptase 85
mast cell derived tryptase 85
matrix metalloproteinase (MMP) 126, 210
metabonomics 150
migration, NE inhibitors 82
migration, neutrophil 84
migration, neutrophil elastase 82
mitogen activated protein kinase (MAPK) 63, 108, 111
MMP/TIMP balance 126
MMP-1 (human collagenase-1) 88, 89
MMP-2 (gelatinase A) 89
MMP-8 (collagenase-2) 89
MMP-9 (gelatinase-B) 88, 89
MT (membrane type) 1-MMP 89
MUC gene 102, 105, 108, 113
MUC5AC 102, 104
MUC5B 102, 104
mucin 101, 104, 113
mucociliary clearance 102, 104, 106
mucolytic agent 109
mucolytic-mucoactive drugs 112
mucoregulation 207
mucous gland hyperplasia 76, 85
mucus 64, 76, 85, 101–106, 113, 207
mucus clearance 104, 106
mucus hypersecretion 101, 102, 104, 105, 113
mucus secretion 64, 76
mucus secretion, ROS 64
Munc18B 111
muscle function 172
myeloperoxidase 61, 122

229

myristoylated alanine-rich C kinase substrate (MARCKS) 108, 110, 112

N-acetylcysteine 109, 112
Nacystelyn 112
NE knockout mouse 84
negative expiratory pressure (NEP) 40
nervous system effect 161
neutrophil 76–86, 107, 108, 128, 130, 211
neutrophil, activation 78, 85
neutrophil, adhesion 78
neutrophil, degranulation 78, 79
neutrophil, migration 78, 79, 84
neutrophil, proteinase 84
neutrophil, proteinase inhibitor 84
neutrophil chemotaxis 130
neutrophil elastase 76, 107, 108
neutrophil elastase, activity 76
neutrophil elastase, bacteria 79
neutrophil elastase, cell surface proteinases 81
neutrophil elastase, Chediak-Higashi syndrome 83
neutrophil elastase, ciliary function 76
neutrophil elastase, epithelial damage 76
neutrophil elastase, exacerbation 78
neutrophil elastase, inhibitor of 76, 78
neutrophil elastase, migration 82
neutrophil elastase, mucous gland hyperplasia 76
neutrophil elastase, mucus 76
neutrophil elastase, source 76
neutrophil elastase, structure 76
neutrophil elastase, synthesis 76
neutrophil elastase, tissue damage 79
neutrophil elastase inhibitor, animal model 86
neutrophil maturation 76
NF-κB inhibitor 206
nitrate 140
nitric oxide (NO) 66, 137, 159
nitrite 140
nitrones 68
nitrotyrosine 140

NO synthase, inducible 109
non-pharmacological intervention 170
nutritional modulation 181
nutritional support 179

obstructive bronchiolitis 192
osteoporosis 161
osteo-skeletal effect 161
out-patient rehabilitation 171
oxidants 107
oxidative damage 140
oxidative stress 157

Papillon-Lefèvre syndrome 84
PDE4 inhibitor 107, 108
peak expiratory flow 33
P-450 enzymes 106
PGE_2 141
$PGF_{2\alpha}$ 141
phosphodiesterase (PDE) 107, 198
phosphodiesterase-4 inhibitor 198
phospholnositide 3-kinase inhibitor 207
p38 MAP kinase inhibitor 206
polymorphism, genetic 89
polymorphism, MMP-1 89
polymorphism, MMP-9 89
polymorphism, MMP-12 89
polymorphism, TIMP-2 89
population admixture 5
PPAR activator 207
P_{2Y2} receptor 113
prostanoid 140
protease 107, 108, 210
protease inhibitor 107, 108
proteinase 3, animal model 85
proteinase 3, emphysema 85
proteinase G 107
proteinase-activated receptor 85, 208
proteinase-activated receptor-2 (PAR-2) 208
proteinase-antiproteinase theory, emphysema 75
proteomics 150

pulmonary function tests, correlation with CT 53
pulmonary rehabilitation program 170
pulmonary vascular disease 197

quality of life 169
quantum proteolysis theory 81

reactive nitrogen species 140
reactive oxygen species (ROS) 61–65
rehabilitation 167
respiratory bronchiolitis-interstitial lung disease (RB-ILD) 54
respiratory muscle function 41
retinoic acid 211
rofecoxib 106
ROS, effects on airways 62–65

secretory leukoproteinase inhibitor (SLPI) 86
sedentarism 158
sensory nerve 110
serine proteinase inhibitor 76
skeletal muscle dysfunction 157
sleep hypoxaemia 42
smoker 140
spin-trap antioxidants (nitrones) 68
sputum 103, 105, 122, 127

sputum eosinophilia 127
sputum induction 122
static pressure volume curve 39
submucosal gland hypertrophy 103
superoxide anion (O_2^-) 61
superoxide dismutase (SOD) 64
surrogate marker 137
systemic extrapulmonary effect 198
systemic impairment 168
systemic inflammation 155

tachykinin 109
thiobarbituric acid reactive substances (TBARS) 66
TIMP-2 89
tiotropium 109
tissue hypoxia 158
tissue inhibitors of metalloproteinases (TIMPs) 88, 126
TNF-α 88, 124
transcription factor nuclear factor-κB (NF-κB) 62, 63, 65–67, 206
transforming growth factor-β1 (TGF-β$_1$) 208
tumour necrosis factor-α (TNF-α) inhibitor 205
tyrosine 140

Wegener's granulomatosis, proteinase 3 85
weight loss, unexplained 157

The PIR-Series
Progress in Inflammation Research

Homepage: http://www.birkhauser.ch

Up-to-date information on the latest developments in the pathology, mechanisms and therapy of inflammatory disease are provided in this monograph series. Areas covered include vascular responses, skin inflammation, pain, neuroinflammation, arthritis cartilage and bone, airways inflammation and asthma, allergy, cytokines and inflammatory mediators, cell signalling, and recent advances in drug therapy. Each volume is edited by acknowledged experts providing succinct overviews on specific topics intended to inform and explain. The series is of interest to academic and industrial biomedical researchers, drug development personnel and rheumatologists, allergists, pathologists, dermatologists and other clinicians requiring regular scientific updates.

Available volumes:
T Cells in Arthritis, P. Miossec, W. van den Berg, G. Firestein (Editors), 1998
Chemokines and Skin, E. Kownatzki, J. Norgauer (Editors), 1998
Medicinal Fatty Acids, J. Kremer (Editor), 1998
Inducible Enzymes in the Inflammatory Response,
    D.A. Willoughby, A. Tomlinson (Editors), 1999
Cytokines in Severe Sepsis and Septic Shock, H. Redl, G. Schlag (Editors), 1999
Fatty Acids and Inflammatory Skin Diseases, J.-M. Schröder (Editor), 1999
Immunomodulatory Agents from Plants, H. Wagner (Editor), 1999
Cytokines and Pain, L. Watkins, S. Maier (Editors), 1999
In Vivo Models of Inflammation, D. Morgan, L. Marshall (Editors), 1999
Pain and Neurogenic Inflammation, S.D. Brain, P. Moore (Editors), 1999
Anti-Inflammatory Drugs in Asthma, A.P. Sampson, M.K. Church (Editors), 1999
Novel Inhibitors of Leukotrienes, G. Folco, B. Samuelsson, R.C. Murphy (Editors), 1999
Vascular Adhesion Molecules and Inflammation, J.D. Pearson (Editor), 1999
Metalloproteinases as Targets for Anti-Inflammatory Drugs,
    K.M.K. Bottomley, D. Bradshaw, J.S. Nixon (Editors), 1999
Free Radicals and Inflammation, P.G. Winyard, D.R. Blake, C.H. Evans (Editors), 1999
Gene Therapy in Inflammatory Diseases, C.H. Evans, P. Robbins (Editors), 2000
New Cytokines as Potential Drugs, S. K. Narula, R. Coffmann (Editors), 2000
High Throughput Screening for Novel Anti-inflammatories, M. Kahn (Editor), 2000
Immunology and Drug Therapy of Atopic Skin Diseases,
    C.A.F. Bruijnzeel-Komen, E.F. Knol (Editors), 2000
Novel Cytokine Inhibitors, G.A. Higgs, B. Henderson (Editors), 2000
Inflammatory Processes. Molecular Mechanisms and Therapeutic Opportunities,
    L.G. Letts, D.W. Morgan (Editors), 2000

Cellular Mechanisms in Airways Inflammation, C. Page, K. Banner, D. Spina (Editors), 2000
Inflammatory and Infectious Basis of Atherosclerosis, J.L. Mehta (Editor), 2001
Muscarinic Receptors in Airways Diseases, J. Zaagsma, H. Meurs, A.F. Roffel (Editors), 2001
TGF-β and Related Cytokines in Inflammation, S.N. Breit, S. Wahl (Editors), 2001
Nitric Oxide and Inflammation, D. Salvemini, T.R. Billiar, Y. Vodovotz (Editors), 2001
Neuroinflammatory Mechanisms in Alzheimer's Disease. Basic and Clinical Research,
    J. Rogers (Editor), 2001
Disease-modifying Therapy in Vasculitides,
    C.G.M. Kallenberg, J.W. Cohen Tervaert (Editors), 2001
Inflammation and Stroke, G.Z. Feuerstein (Editor), 2001
NMDA Antagonists as Potential Analgesic Drugs,
    D.J.S. Sirinathsinghji, R.G. Hill (Editors), 2002
Migraine: A Neuroinflammatory Disease? E.L.H. Spierings, M. Sanchez del Rio (Editors), 2002
Mechanisms and Mediators of Neuropathic pain, A.B. Malmberg, S.R. Chaplan (Editors),
    2002
Bone Morphogenetic Proteins. From Laboratory to Clinical Practice,
    S. Vukicevic, K.T. Sampath (Editors), 2002
The Hereditary Basis of Allergic Diseases, J. Holloway, S. Holgate (Editors), 2002
Inflammation and Cardiac Diseases, G.Z. Feuerstein, P. Libby, D.L. Mann (Editors), 2003
Mind over Matter – Regulation of Peripheral Inflammation by the CNS,
    M. Schäfer, C. Stein (Editors), 2003
Heat Shock Proteins and Inflammation, W. van Eden (Editor), 2003
Pharmacotherapy of Gastrointestinal Inflammation, A. Guglietta (Editor), 2004
Arachidonate Remodeling and Inflammation, A.N. Fonteh, R.L. Wykle (Editors), 2004